ENGINEERING DESIGN GRAPHICS

Combination Text and Problems

James H. Earle

Texas A&M University

Second printing, July 2014 (7CX6.31)
First printing, July 2012 (1.5X00.7)
Copyrighted by Creative Publishing
ISBN-10: 0615671233
ISBN-13: 978-0-615-67123-9

Creative Publishing
P.O. Box 9292, PH 800-245-5841
College Station, Texas 77842

FAX 979-764-7758
creativepublishing@verizon.net
www.creativepublishing.com

To the Student

This book is a combination (combo) of explanatory text accompanied by related problems to aid you in learning and applying the concepts covered by the text. The problems will be assigned, graded, and returned by your instructor for future reference and review for periodic tests.

Notebook: You should acquire a two-inch thick, three-ring notebook that will become your portable file in which this entire book will be inserted at the beginning of the course. Additional details on the use of your notebook are given on page 7.

Presentation of Content: Explanatory examples, illustrations, and supporting text are included on each problem page, enabling you to solve similar problems on the same page adjacent to the examples. This close association of instructions with the problems will be more convenient than referring to the illustrations in a separate textbook. Problems will progress from very fundamental to more advanced to aid in your learning process.

Introduction to Design: An overview of the design process is given in pages 8 through 28 to illustrate the recommended steps of design and the importance of graphics as the primary tool of the creative process. Remember that creativity is enhanced and communicated by graphics from the first sketch of a concept to its final presentation as a working drawing from which a design becomes reality.

Computer Graphics Problems: Computer problems similar to those on front of the pages are given on the back of most sheets with space for your solution by computer. The first computer problem is given on page 52B (the back of sheet 52). Specific computer instructions have been omitted due to the variation in computer software that may be used. Your instructor will provide instruction in the use of the software available in your program.

Applications: Many real problems are given that can be solved with graphics and design principles as would be encountered in industry. Other design problems with multiple solutions are given throughout this publication.

Grades: Record the grades for each problem sheet, test, and assignment on page 4 that will enable you to follow your progress throughout the course. File your graded sheets and assignments in your three-ring notebook for future reference as described above. The percentages for the various types of assignments listed below will be given to you by your instructor.

Number

. . . .	Problem sheets	. . . . . %
. . . .	Periodic tests	. . . . . %
. . . . .	Design project	. . . . . %
. . . . .	Working drawing 1	. . . . . %
. . . . .	Working drawing 2	. . . . . %
. . . . .	Optional assignments	. . . . . %
. . . . .	Other	. . . . . %
	TOTAL	. . . . . %

Table of Contents

Summary

The sequence of the course content listed above may be modified by your instructor. Strive to develop your skills in the development, application, and presentation of your design concepts through the application of graphics presented in this course. Retain your problem sheets throughout the course in a three-ring notebook. After the assigned problem sheets have been completed, graded, and returned by your teacher, return them to the three-ring notebook as a permanent reference during and after the course. Your notebook, with its completed and graded problem sheets, will serve as an example of your knowledge of graphics when you apply for a job.

When you enter the field of technology or engineering, you will be involved in the preparation of drawings in the development and presentation of designs and projects throughout your career. Graphics will be a means of developing, refining, and presenting new ideas for new products, and their development.

Always remember that graphics is the language of design.

Grade Sheet

Probs.	Grade

Probs.	Grade

Weekly Test Grades

Test 1 _____   Test 8 _____

Test 2 _____   Test 9 _____

Test 3 _____   Test 10 _____

Test 4 _____   Test 11 _____

Test 5 _____   Test 12 _____

Test 6 _____   Test 13 _____

Test 7 _____   Test 14 _____

Test Avg: _____

Design Project

Problem Identification (. . . %) _____

Preliminary Ideas (. . . %) _____

Refinement (. . . %) _____

Analysis (. . . %) _____

Decision & Finalization (. . . %) _____

Presentation (. . . %) _____

Working Drawings

Working Drawing 1 (. . . %) _____

Working Drawing 2 (. . . %) _____

Daily Grades

Problem Sheets (. . . %) _____

Tests

Test Grades (. . . %) _____

Final Exam (. . . %) _____

Final Grade (100%) _____

Working Drawings

Team No. _____   Project:_____

	Names	No. (N=)	% Contribution (C	F=NC	GRADE
1.	_____		_____	_____	_____
2.	_____		_____	_____	_____
3.	_____		_____	_____	_____
4.	_____		_____	_____	_____
5.	_____		_____	_____	_____
6.	_____		_____	_____	_____
7.	_____		_____	_____	_____
8.	_____		_____	_____	_____

Evaluation by your Instructor

TITLE BLOCK (5 points)

Student's name	1	. . . .
Checker's name	1	. . . .
Date	1	. . . .
Scale	1	. . . .
Sheet number	1	. . . .

DRAWING DETAILS (19 pts)

Properly-drawn views	10	. . . .
Spacing of views	5	. . . .
Part names and numbers	2	. . . .
Sections and conventions	2	. . . .

DESIGN INFORMATION (12 pts)

Proper tolerances	5	. . . .
Surface texture symbols	2	. . . .
Thread notes and symbols	5	. . . .

DIMEN. PRACTICES (20 pts)

Proper arrowheads	3	. . . .
Positions of dimensions	3	. . . .
Completeness of dimens.	4	. . . .
Fillet & round notes	2	. . . .
Machined-hole notes	2	. . . .
Inch/mm marks omitted	2	. . . .

Dimensions in best views	4	. . . .

DRAFTSMANSHIP (19 pts)

Thicknesses of lines	6	. . . .
Lettering	6	. . . .
Neatness	3	. . . .
Reproduction quality	4	. . . .

GENERAL NOTES (5 pts)

SI symbol	2	. . . .
Third angle symbol	2	. . . .
General tolerance note	1	. . . .

ASSEMBLY DRAWING (15 pts)

Descriptive views	6	. . . .
Clarity of assembly	3	. . . .
Parts list completness	4	. . . .
Part numbers in balloons	2	. . . .

PRESENTATION (5 pts)

Properly stapled	2	. . . .
Properly trimmed	1	. . . .
Properly folded	1	. . . .
Grade sheet attached	1	. . . .

Total 100 . . . .

Oral Report

The table below should be completed jointly by the team with only the grade column completed by the instructor who will use the chart on page 4 and the factor F that was computed for each member.

Team No. _____ Project: _____

Names	No. (N=)	%Contri-bution (C)	F=NC	GRADE
1. _____		_____	_____	_____
2. _____		_____	_____	_____
3. _____		_____	_____	_____
4. _____		_____	_____	_____
5. _____		_____	_____	_____
6. _____		_____	_____	_____
7. _____		_____	_____	_____

Evaluation by the instructor   Max.   Comments:

		Max.	
1.	Introduction of team members	2	. . . .
2.	Proper dress of team members	2	. . . .
3.	Statement of purpose of presentation	5	. . . .
4.	Use of visuals—point to important points, do not block screen, do not fumble, etc.	10	. . . .
5.	Adequate number of visual aids	9	. . . .
6.	Quality of visual aids	15	. . . .
7.	Clear presentation of recommended design	10	. . . .
8.	Presentation of alternate solutions	2	. . . .
9.	Consideration of human factors	5	. . . .
10.	Coverage of economics (manufacturing, shipping, packing, overhead, markup, etc.)	10	. . . .
11.	Presentation of an effective conclusion	5	. . . .
12.	Continuity of presentation	3	. . . .
13.	Poise and professionalism	2	. . . .
14.	Team participation (perfect score if all participate)	10	. . . .
15.	Use of allotted time	10	. . . .

TOTAL 100   . . . .

Other instructor comments on back of this sheet.

6

Record Keeping

Using This Book:
The pages of this textbook-problem book combination (referred to as the *Combo*) will enable you to learn and apply the various principles of graphics and design with a single book. The textbook principles needed to solve the problems are located on each page adjacent to the problems, making reference to the instructions both convenient and instantaneous.

Notebook Filing:
It is recommended that you insert the entire text/problem book, cover and all, into a three-ring notebook with two-inch diameter rings at the beginning of the course as a permanent file. Sheets can be removed one at a time as they are assigned, solved, turned in and graded. Graded sheets should be replaced in their original sequence in the three-ring notebook after being returned by your instructor. This reinsertion of graded sheets will protect and preserve them throughout the course, making them available for reference during and after completion of the course.

Review for Tests:
By retaining the problem sheets before and after completion as described above, your notebook and its contents will serve as a source book for periodic review for tests and references.

Tabulate Your Grades:
List your grades when your problem sheets are returned on the *Grade Sheet* on page 4. You should be the first to know what your grades are in this course as it progresses.

Design Projects:
Design problems are suggested on pages 27 and 28 from which to select your project as described in *The Design Process* unit, pages 8-24. (Perhaps you have a design problem to propose to your instructor instead of one from pages 27 and 28.)

The Design Process example on pages 8-24 illustrates how the *Design Project* sheets on pages 29-44 were used in designing an exercise device. Included on pages 20-24 are detailed drawings, specifications, and photographs of the final product.

A-size (8.5 x 11) and B-size (11 x 17) sheets will be adequate for these drawings as illustrated on pages 20-22. B-size sheets are given at the rear of the book for your design project.

Future Reference:
Since the material covered here will apply to other courses and career applications, think of your notebook as a reference source that can be used as an example of your background when applying for a job. Your skills with pencil and computer, coupled with the application of graphics, will be used throughout your career.

Remember that graphics
is the language of design.

UNIT 1

The Design Process

1.1 Introduction

The design process is the method of creating innovative solutions to problems that will result in new products or systems. Engineering graphics is the primary medium of design that is used for developing designs from initial concepts to final working drawings. Initially, a design consists of sketches which are refined, analyzed, and developed into precise detail drawings and specifications. They, in turn, become part of the contract documents for the parties involved in funding and implementing a project.

At first glance, the solution of a design problem may appear to involve merely the recognition of a need and the application of effort toward its solution, but most engineering designs are more complex than that. The engineering and design efforts may be the easiest parts of a project **(Figure 1.1)**.

For example, engineers who develop roadway systems must deal with constraints such as ordinances, historical data, human factors, social considerations, scientific principles, budgeting, and politics. Engineers can readily design driving surfaces, drainage systems, overpasses, and other components of the system. However, adherence to budgetary limitations is essential, and funding is closely related to politics on public projects.

Traffic laws, zoning ordinances, environmental impact statements, right-of-way acquisition, and liability clearances are legal as-

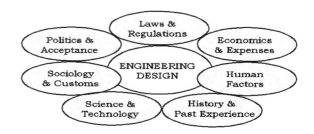

1.1: An engineering project often involves the interaction of people representing many professions and interests, with engineering design as the central function.

pects of roadway design that engineers must deal with. Past trends, historical data, human factors (including driver characteristics), and safety features affecting the function of the traffic system must be analyzed. Social problems may arise if proposed roadways will be heavily traveled and will attract commercial development such as shopping centers, fast-food outlets, and service stations. Finally, designers must apply engineering principles developed through research and experience to obtain durable roads, economical bridges, and fully functional systems.

1.2 Terminology

The content of this book has been developed to enable you to learn the principles of graphics necessary for the preparation of drawings

8

that will bring ideas into reality. Graphics, whether done by hand or by computer, is an essential medium from the beginning step through the final step in creating products or systems.

Creativity: Creativity can manifest itself in limitless ways ranging from unique ideas applied to everyday situations to highly complex scientific innovations. However, creativity is not functional and lies dormant as an unexploited clever concept unless it is applied, thereby launching the process of design.

Artistic Design: Design may be applied to the creation of a two-dimensional painting, a three-dimensional sculpture, or other non-utilitarian applications. The resulting artistic products are simply meant to be attractive, decorative, and pleasing to the eye but are seldom functional.

Technical Design: The application of creativity applied to the development of a set of drawings and specifications that can be used to produce products or systems is technical design. A degree of artistic design is often part of the solution as well, but only after the functional aspects of the design are solved.

Drafting/Graphics: The step in which designs are presented in final form as working drawings using the appropriate standards and details is referred to as drafting or engineering drawing. Much of this process is less creative than the previous steps; it is simply that part of the process where experience, knowledge, and skills are applied finalize and document the design.

1.3 Creativity, Design, and You

Have you ever recognized a condition that needed a solution? That's the first step of design and of being a designer. Your concept need not be a major earth-shaking idea for it to be worthwhile and profitable, it just needs to fill a need. The old-fashioned manual key was OK, but the punch-button device of today is much better and more convenient **(Figure 1.2)**. The first version had an ignition key that was separate from the actuation device. Today's key is a one-piece combination of the two operations merged into a single device.

What about a tab of paper that is sticky on

1.2: What's more convenient than a device for locking and unlocking a car door from a distance. Maybe a key with this built-in capability?

1.3: How about a "sticky" to use as a book mark? What could be a simpler solution to fill this need? We let another good idea get away.

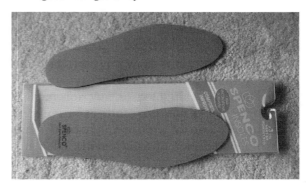

2.4: Who doesn't have uncomfortable shoes from time to time? Insoles that can be inserted in your exisiting shoes may be just the solution for you.

one side to make the tab stick to page as a marker? A simple solution, but a big money maker that is widely used **(Figure 1.3)**.

Another product that should have been easy to think of are shoe inserts for making your shoes more comfortable **(Figure 1.4)**. Couldn't you have thought of them as marketable products?

1.5: A breakthrough in beverage cans was the snap-open top that eliminated the need for various types of openers that has sinse become the universal standard. *ALCOA, www.alcoa.com.*

1.6: Chrysler's Peapod has a maximum speed of 25 mph at a cost of two cents per mile after plugging into a 110-volt outlet. It is designed for running errands but not for highway driving. *peapodmobility.com*

Product Design Characteristics

1.7: Product design seeks to develop a product that meets a specific need, that can function independently, and that can be mass produced.

Then there's the snap-top beverage can that has become the industry standard by removing the need for openers **(Figure 1.5).**

These were problems whose solutions were not desperately needed, but they are excellent examples of how design opportunities surround us. A more sophisticated design solution is the Peapod automobile that was devel-

Factors of Product Design

1.8: These are some of the major factors that must be considered when developing a product design.

SYSTEMS DESIGN
Entertainment
Water provisions
Air conditioning
Heating furnace
Lighting services
Gas utilities
Car protection
Food preparation
Bathing facilities
Appliances
Sewage services

1.9: The typical residence is a system composed of many components and products.

oped to fill the mandated need of conserving energy while providing an efficient source of transportation in the highly competitive automobile market **(Figure 1.6).** Do not assume that being creative and designing worthwhile products and systems are beyond your capabilities. The challenge is to learn to use your creativity that begins with curiosity and a recognition of problems in need of solution.

I Could Have Done That! Just maybe, this textbook and the course in which it is used will encourage you to reach for opportunity, to apply the principles of creativity, and to achieve success that is limited only by your imagination. And above all, never allow yourself to be in that position where you will say, "I could have done that!"

1.4 Types of Design Problems
Most design problems fall into one of two categories: **product design** and **systems design.**

Product Design
Product design is the creation, testing, and manufacture of an item that usually will be mass produced, such as appli-

SMARTE CARTE: A PRODUCT AND PART OF A SYSTEM

This luggage cart is a product and part of a coin-operated luggage system in airports for the convenience of travelers in managing their baggage.

1.10: The Smarte Carte® is a product and also part of a system. *Smarte Carte.* www.smartecarte.com

SMARTE CARTE STATION

Luggage stations are located near entry gates and baggage areas where individual carts can be returned after use and dispensed to other travelers. This combination of cart and a dispenser is the beginning of a system which offers more than just a product.

1.11: Carts plus a dispensing apparatus is the beginning of the development of a system. (Courtesy of Smarte Carte®.)

COIN-OPERATED DISPENSER

In addition to releasing a cart, the station makes change, accepts credit cards, and issues baggage cards that can be used at the next airport during the traveler's trip.
These added features comprise a system that is more than just a grouping of products.

1.12: The coin-operated dispenser station releases a cart, makes change for $5, $10, and $20 bills, and issues baggage cards that can be used at the next terminal. (Courtesy of Smarte Carte®.)

ances, tools, or large products such as automobiles (Figure 1.7). In general, a product must have sufficiently broad appeal for meeting a specific need and performing an independent function to warrant its production in quantity. Designers of products, whether an automobile or a bicycle, must consider current market needs, production costs, function, sales, distribution methods, profit predictions, and other factors shown in **Figure 1.8.**

Products can perform one or many functions. For instance, the primary function of an automobile is to provide transportation, but it also contains products that provide communications, illumination, comfort, entertainment, and safety. Because it is mass produced for a large consumer market and can be purchased as a unit, the automobile is regarded as a product. However, because it consists of many products that perform various functions, the automobile is also a system.

Systems Design

Systems design combines products and their components into a unique arrangement and provides a method for their operation. A residential building is a system of products consisting of heating and cooling, plumbing, natural gas, electrical power, sewage handling, appliances, entertainment, and others that form the overall system as shown in **Figure 1.9.**

Systems Design Example

Suppose that you were carrying luggage to a faraway gate in an airport terminal. It would be easy for you to recognize the need for a luggage cart that you could use and then leave behind for others to use. If you have this need, then others do, too. The identification of this need could prompt you to design a cart like the one shown in **Figure 1.10** to hold luggage, and even a child, and to make it available to travelers. The cart is a product.

How could you profit from providing such a cart? First you would need a method of holding the carts and dispensing them to customers, such as the one shown in **Figure 1.11.** You would also need a method for users to pay for cart rental, so you could design a coin-operated gate for releasing them (**Figure 1.12**).

For an added customer convenience, you could provide a vending machine that will take

1.13: An example of a complex system is the monorail network that is composed of vehicles, rails, bridges, signals, and so forth.

The Design Process

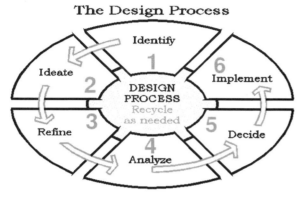

1.14: The design process consists of six steps, each of which can be recycled as needed.

bills both to make change and to issue cards entitling customers to multiple use of carts at this and other terminals (**Figure 1.12**). A mechanism to encourage customers to return carts to another conveniently located dispensing unit would be helpful and efficient. A coin dispenser (**Figure 1.11**) that gives a partial refund on the rental fee to customers returning carts to a dispensing unit at their destination might work.

The combination of these products and the method of using them is a system design, although it is a relatively simple one. Such a system of products is more valuable than the sum of the products alone. An example of a larger, more complex system is the monorail system (**Figure 1.13**) that is composed of vehicles, elevated rails, bridges, tunnels, signals, signs and much more.

1.5 The Design Process

Design is the process of creating a product or system to satisfy a set of requirements that has multiple solutions by us-

1.15: Problem identification requires that the designer accumulate as much information about a problem as possible before attempting a solution. The designer also should keep product marketing in mind at all times.

ing many available resources. In essentially all cases, the final design must be profitable or within a budget.

The steps of the design process shown in **Figure 1.14** are:

1. problem identification,
2. preliminary ideas (ideation),
3. refinement,
4. analysis,
5. decision, and
6. implementation.

Designers should work sequentially from step to step but should review previous steps periodically and rework them if a new approach comes to mind during the process.

Problem Identification

Most engineering problems are not clearly defined at the outset and require identification before an attempt is made to solve them (**Figure 1.15**). For example, air pollution is a concern, but we must identify its causes before we can solve the problem. Is it caused by automobiles, factories, atmospheric conditions that harbor impurities, or geographic features that trap impure atmospheres?

Another example is traffic congestion. When you enter a street where traffic is unusually congested, can you identify the reasons for the congestion? Are there too many cars? Are the signals poorly synchronized? Are there visual obstructions? Has an accident blocked traffic?

Problem identification involves much more than simply stating, "We need to eliminate air pollution." We need data of several types: opinion surveys, historical records, personal observations, experimental data, physical

Factors of Preliminary Ideas

1.16: The designer gathers ideas from a brainstorming session and develops preliminary ideas for problem solution. Ideas should be listed, sketched, and noted to have a broad range of ideas to work with.

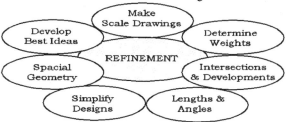

Factors of Preliminary Ideas

1.17: Refinement begins with the construction of scale drawings of the best preliminary ideas. Descriptive geometry and graphics are used to describe geometric characteristics.

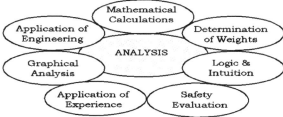

Design Analysis

1.18: All available methods, from science to technology to graphics to experience, should be used to analyze a design.

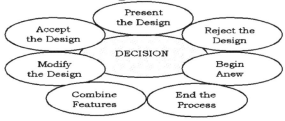

Making a Decision

1.19: Decision involves the selection of the best design or design features to implement. This step may require an acceptance, rejection, or a compromise of the proposed solution.

measurements from the field, and more. It is important that the designer resist the temptation to begin developing a solution before the identification step has been completed.

Preliminary Ideas

The second step of the design process is the development of as many ideas for problem solution as possible (**Figure 1.16**). A brainstorming session is a good way to collect ideas at the outset that are highly creative, revolutionary, and even wild. Rough sketches, notes, and comments can capture and preserve preliminary ideas for further refinement. The more ideas, the better at this stage.

Refinement

Several of the better preliminary ideas are selected for refinement to determine their merits. The rough preliminary sketches are converted into scale drawings for spatial analysis, determination of critical measurements, and the calculation of areas and volumes affecting the design (**Figure 1.17**). Descriptive geometry aids in determining spatial relationships, angles between planes, lengths of structural members, intersections of surfaces and planes, and other geometric relationships.

Analysis

Analysis is the step during which engineering and scientific principles are used most intensively to evaluate the best designs and compare their merits with respect to function, strength, safety, cost, and optimization (**Figure 1.18**). Graphical methods play an important role in analysis, also. Data can be ana-

lyzed graphically, forces analyzed as graphical vectors, and empirical data can be analyzed, integrated, and differentiated by other graphical methods. Analysis is less creative than the previous steps.

Decision

After analysis, a single design, which may be a compromise among several designs, is decided upon as the solution to the problem (**Figure 1.19**). The designer alone, or a team, may make the decision. The outstanding aspects of each design usually lend themselves to graphical comparisons of manufacturing costs,

Implementation Step

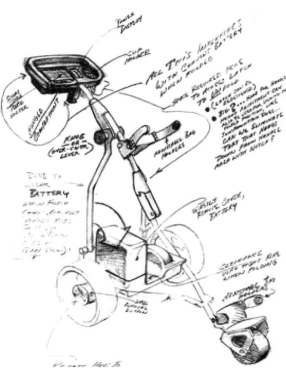

1.20: Implementation is the preparation of drawings, specifications, and documentation from which the product can be made. The product is produced and marketing is begun.

weights, operational characteristics, and other data essential in decision making.

Implementation

The final design must be described and detailed in working drawings and specifications from which the project will be built, whether it is a computer chip or a suspension bridge (**Figure 1.20**). Workers must have precise instructions for the manufacture of each component, often measured within thousandths of an inch to ensure proper fabrication assembly. Working drawings must be sufficiently explicit to serve as part of the legal contract with the successful bidder on the job.

1.6 Graphics and Design

Whether by freehand sketching, with instruments, or on a computer, **graphics is the medium of design.** Engineering, scientific, and analytical principles must be applied throughout the process, and graphics is the medium that is employed in each step from problem identification to implementation.

An example of how graphics is used as a medium of design is the development of the Hawk International Corporation's electrically powered Golf Caddy, which was designed by Cesaroni Design Associates, Inc. shown in **Figures 1.21** through **Figure 1.25.** The designer makes numerous rough freehand pencil sketches of design ideas, using paper and pencil along with notes as shown in **Figure 1.21.** Many of these design sketches are made in an evolutionary manner to enable ideas to evolve and develop during the process.

The designer makes a series of three-dimensional sketches of the caddy from his previous concepts, two of which are shown in **Figures 1.22 and 1.23.** These drawings repre-

1.21: The designer makes numerous freehand sketches to develop his ideas for a motorized golf-bag cart and describe them to others. *(Cesaroni Design Associates, Inc., www.cesaroni.com)*

1.22: The better concepts are more carefully sketched and detailed as more realistic renderings to enhance their evaluation. *(Cesaroni Design Associates, Inc.)*

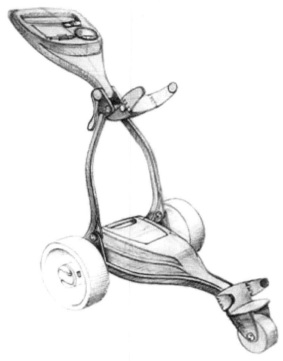

1.23: Styling is also considered during the early stages of concept development. It is an important aspect of almost all products to make them more marketable, but function is the most important factor of all. *(Cesaroni Design Associates, Inc.)*

1.24: It should be no surprise that the end product looks like the designer's sketches that were developed earlier. *(Cesaroni Design Associates, Inc.)*

1.25: Being able to fold the caddy to fit in the trunk of the golfer's car is an important feature that contributes to its marketability. *(Cesaroni Design Associates, Inc.)*

sent the designer's thinking process as well as a means of recording his ideas and for communicating with others. But most of all, this is the manner in which the **designer communicates with himself** as he progresses toward a solution.

Additional three-dimensional pictorials are drawn to develop and illustrate necessary operational details as the design evolves. The preliminary designs are studied and evaluated by several designers as a means of "trouble shooting" the design and applying all available experience and expertise to the design.

Engineering must be applied as the design progresses toward its conclusion to assure that the electronics can be properly housed, that manufacturing can be economically performed, and that it functions in the best manner. Scale models made for analysis are developed into full-size prototypes as the final solution is approached. These steps enable the caddy to be evaluated on the basis of its function and its styling.

After approval, the design becomes a reality as shown in **Figure 1.24.** Since it is a portable device, it was designed to fold up to a size that would permit it to be carried in the trunk of the golfer's car **(Figure 1.25).** This example illustrates only a few of the high points of the design process in arriving at its solution. The following case study of the Iron Gym® offers a more thorough coverage of the application of the design process.

1.7 Case Study: Design Process
The following example illustrates the application of the design process to a simple problem with no electronics, motors, complicated devices.

Home Exerciser

Your organization is in need of a product that could be offered to the general public that will be beneficial and therefore profitable. The steps of the design process have been applied to this need in the following example to illustrate how a real product was brought into reality.

Problem Identification

Most designs begin with a blank sheet of paper, a pencil, and a few vague ideas and little more. During most of the process, the designer communicates with himself as this case study will show. Let's begin by writing a statement of the problem and listing its general requirements and limitations (**Sheet 1**). Almost immediately you will realize that you need additional information and data even before the project begins. List needed information and market considerations along with possible sources where this data might be found in your follow-up search for these answers (**Sheet 2**). Even if some of this information may be obvious, writing statements and making notes about the problem will help you "warm up" to the problem and begin the creative process. Also, you must begin thinking about sales outlets and marketing methods without which your product will be a failure.

Preliminary Ideas

Brainstorm the problem for possible directions or solutions with your associates, or alone by yourself if necessary. (Brainstorming will be applied to problems on **Sheets 27** and **28**.) List the ideas and concepts that were obtained on a worksheet for your permanent file (**Sheet 3**), and summarize the best ideas and design features on a separate worksheet (**Sheet 4**). Then translate and expand these verbal ideas into rapidly drawn freehand sketches along with explanatory notes (**Sheet 5** and **Sheet 6**). Develop as many ideas as possible during this step because a large number of ideas represents a high level of creativity. This is the most creative step of the design process, and there are few restrictions and limitations at this stage. Preserve ALL your ideas and concepts for reference as the design process progresses. **The designer is mostly communicating with himself at this stage of the design process.**

PROBLEM IDENTIFICATION

1. Project title
   HOME EXERCISER

2. Problem statement
   DESIGN A DEVICE THAT WOULD ENHANCE HOME EXERCISING FOR THE GENERAL MARKET.

3. Requirements and limitations
   A. SUPPORT SEVERAL EXERCISES
   B. RETAIL PRICE $25–$45 RANGE
   C. BETTER IF FREE-STANDING, AND DOES NOT ATTACH TO WALL
   D. SMALL AND EASY TO STORE UNDER BED OR ON SHELF
   E. BETTER IF NO WEIGHTS WERE REQUIRED FOR LIFTING
   F. SUITABLE FOR MARKETING BY MAIL, LIGHTWEIGHT SHIPPING
   G. EASILY BOXED FOR SHIPPING,
   H. SIMPLE TO USE

| PROBLEM IDENTIFICATION | NAME NO. SECT DATE | TIME | 1 |

Sheet 1: The problem identification worksheet itemizes various factors that must be considered at the beginning of a products's design.

Sheet 2: A continuation of problem identification is shown on this worksheet to list needed information and possible sources from where it can be obtained.

---

4. Needed information
   A. EXERCISE THAT IS MOST NEEDED? WHAT MUSCLES ARE WEAKEST? —CHECK MEDICAL MANUALS
   B. THE DEVICES MOST USED? —REVIEW PRODUCTS ON THE INTERNET AND IN MAGAZINES
   C. IS THERE A UNIQUE DEVICE ON MARKET NOW? —CHECK CATALOGS
   D. WHAT ARE SIZES OF MOST ON MARKET? —CHECK CATALOGS

5. Market considerations
   A. EQUIPMENT BEIING USED NOW? —CHECK COMPETITORS' CATALOGS
   B. WHAT IS MARKET POTENTIAL? —CHECK WITH COMPETITORS —INTERVIEW CUSTOMERS —CHECK WITH COMPANY REPS
   C. WHAT IS THE PRICE RANGE? —CHECK WITH OUR FIELD REPS —SURVEY INDIVIDUALS

| PROBLEM IDENTIFICATION | NAME NO. SECT DATE | TIME | 2 |

PRELIMINARY IDEAS

1. Brainstorming ideas

   A. PUSH—UPS DEVICE
   B. LIFTING APPARATUS
   C. USE ELASTIC STRAPS
   D. HANG FROM CEILIING
   E. ADD WATER AS WEIGHT
   F. LIGHTWEIGHT DUMBBELLS
   G. RIG WITH PULLEY SYSTEM
   H. SUPPLEMENT WITH TV TAPE
   I. SIT—UP EXERCISING DEVICE
   J. EXERCISE FOR ENDURANCE
   K. EXERCISE FOR STRENGTH
   L. STRENGTHEN UPPER BODY
   M. STRENGTHEN LOWER BODY
   N. LIFT BODY NOT WEIGHTS
   O. DESIGN TO BE LIGHTWEIGHT
   P. USABLE AT OFFICE
   Q. BEST USED AT HOME
   R. DEVELOP ARM MUSCLES MOST
   S. DEVELOP LEGS AND ARMS
   T. MINIMUM OF MOVING PARTS
   U. STORE UNDER BED

| PRELIMINARY IDEAS | NAME NO. SECT DATE | TIME | 3 |

Sheet 3: Ideas from a brainstorming session are recorded by a member of the brainstorming team as they emerge.

Sheet 4: The ideas most worthy of development are selected from the brainstorming session list to be developed as preliminary ideas.

2. Description of best ideas

   A. UPPER BODY DEVELOPMENT:
      STENGTHEN ARMS FROM
      CHINNING EXERCISES
      MUST SUSPEND OVERHEAD

   B. ADDITIONAL DEVELOPMENT:
      IMPROVE ABS BY PUSH UPS
      AND FROM A SUSPENSION
      POSITION LIKE IN ITEM A

   C. MODIFY DESIGN FOR PUSH UPS
      FROM THE FLOOR

   D. PERHAPS CONNECT TO DOOR
      FRAME OR GARAGE WALL

   E. CONTINUE DEVELOPMENT

3. Attach sketches

| PRELIMINARY IDEAS | NAME NO. SECT DATE | TIME | 4 |

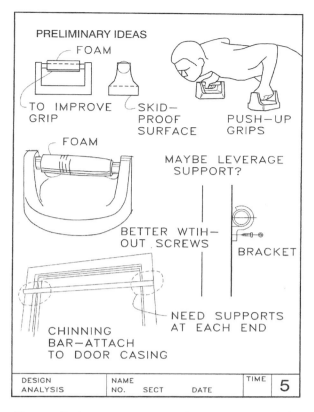

PRELIMINARY IDEAS

FOAM — TO IMPROVE GRIP

SKID—PROOF SURFACE

PUSH—UP GRIPS

FOAM

MAYBE LEVERAGE SUPPORT?

BETTER WTIHOUT SCREWS

BRACKET

CHINNING BAR—ATTACH TO DOOR CASING

NEED SUPPORTS AT EACH END

| DESIGN ANALYSIS | NAME NO. SECT DATE | TIME | 5 |

Sheet 5 Preliminary ideas are sketched and noted for further development. This is the most creative step of the design process.

Sheet 6 Additional preliminary design solutions are sketched and developed here.

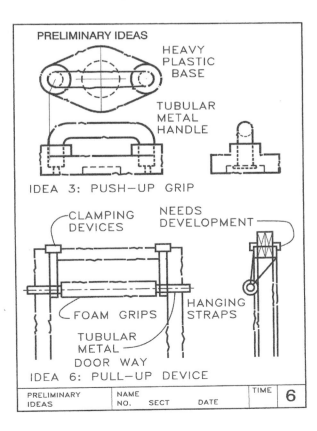

PRELIMINARY IDEAS

HEAVY PLASTIC BASE

TUBULAR METAL HANDLE

IDEA 3: PUSH—UP GRIP

CLAMPING DEVICES

NEEDS DEVELOPMENT

FOAM GRIPS

HANGING STRAPS

TUBULAR METAL DOOR WAY

IDEA 6: PULL—UP DEVICE

| PRELIMINARY IDEAS | NAME NO. SECT DATE | TIME | 6 |

```
REFINEMENT
1. Description of best ideas
   A. OVERHEAD—SUSPENSION DEVICE
      1. DEVICE MADE TO REST ON
         THE UPPER DOOR CASING
      2. NOT PERMANENTLY ATTACHED,
         REMOVABLE AFTER USE
      3. USED FOR CHINNING AND
         DEVELOPMENT OF ABS
      4. DEVELOPS ARMS AND BACK
      4. NO MOVING PARTS

   B. PUSH—UP HANDLES
      1. BENT TUBULAR METAL
      2. LIMITED TO PUSH UPS, NO
         CHINNING OR PULL UPS ARE
         SUPPORTED
      3. SMALL AND LIGHT WEIGHT,
         LIMITED TO ONE EXERCISE
      4. ECONOMICAL SHIPPING, EASY
         TO STORE BY CUSTOMER

2. Attach scale drawings
```

DESIGN REFINEMENT	NAME NO. SECT DATE	TIME	7

Sheet 7: Refinement of preliminary ideas begins with written descriptions of the better ideas.

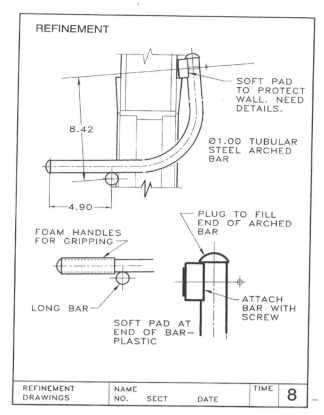

REFINEMENT DRAWINGS	NAME NO. SECT DATE	TIME	8

Sheet 8: Scale drawings of design details are made to describe and develop them. Almost no dimensions are needed.

Sheet 9: The first worksheet of analysis step moves from freedom to development into and investigation of product function and marketability.

Problem Refinement

Describe the design features of one or more preliminary ideas on a worksheet for comparison as shown in **Sheet 7**. Draw the better designs to scale in preparation for analysis; only a few dimensions are needed at this stage (**Sheet 8**).

Use instrument-drawn orthographic projections, computer drawings, and descriptive geometry to refine the designs and ensure precision. Let's say that you select ideas 3 and 6 for analysis. **Sheet 8** depicts orthographic views of the refinement of idea 6.

Analysis

Work sheets are used to analyze the pull-up bar design both verbally and graphically in **Sheets 9** through **12**. If you are still considering more than one solution at this point, analyze each design separately for comparison.

By using the maximum load of 220 lb and the geometry of the bracket, it is possible to graphically determine the angle of the reaction that the bolt must carry, XXX lb, when the bracket is fully loaded (**Sheet 12**). Again, graphics is used as an important design tool.

```
ANALYSIS
1. Function
   A. SUPPORTED BY DOOR FRAME
      WITH SCREWS OR BOLTS
   B. SUPPORTS PERSON DURING THE
      PROCESS OF PULL UPS
   C. SUPPORTS LEG LIFTS
   D. ALSO USED AS FLOOR SUPPORT
2. Human engineering
   A. NO PERMANENT INSTALLATION
   B. REQUIRES NO SPECIAL TOOLS
   C. PROVIDES ADEQUATE BODY
      SUPPORT WHILE IN USE
   D. SAFE AND SUPPORTIVE WHILE
      IN USE
3. Market and consumer acceptance
   A. KEEP PRICE UNDER $40
   B. CAN BE SOLD AS AN ADDITION
      TO AN EXERCISE PRODUCT LINE
   C. COULD BE SOLD AS AN INTERNET
      PRODUCT BY OTHERS
   D. BOX IN A COLORFUL ILLUSTRATED
      CARTON AT RETAIL OUTLETS
```

DESIGN ANALYSIS	NAME NO. SECT DATE	TIME	9

4. Physical description
    A. DOOR DEVICE: DOORS UP TO
       32" WIDE WITH 3.5" CASING
    B. DURABLE FRAME OF TUBULAR
       STEEL & RUBBER SLEEVES
    C. WEIGHT: ABOUT 6-7 POUNDS
    D. OVERALL DIMENSIONS: 33" X
       9" X 4"
    E. FORMED BY A CONNECTION OF
       TUBULAR STEEL MEMBERS

5. Strength
    A. WILL SUPPORT UP TO 300 LBS
       WITH A SAFETY FACTOR OF 1.5
    B. TUBULAR MEMBERS ARE CON-
       NECTED WITH BOLTS AND NUTS
       OF MORE THAN ADEQUATE
       STRENGTH
    C. SUPPORTED BY DOOR CASING
       AND THE LEVERAGE OF THE
       WEIIGHT APPLIED BY THE USER

| DESIGN ANALYSIS | NAME NO.   SECT   DATE | TIME | 10 |

Sheet 10: This worksheet is a continuation of the analysis of the preliminary design options begun in the previous sheet.

Sheet 11: Production and economic considerations are major considerations of the analysis step of the design process and must be studied thoroughly.

6. Production procedures

PULL-UP DEVICE:
    A. FABRICATE BY CONNECTING
       WITH NUTS AND BOLTS
    B. NO WELDIING OF PARTS
    C. RUBBER CYLINDERS APPLIED TO
       STEEL TUBING FOR IMPROVED
       GRIPPING
    D. PAINTED TO GIVE RUST-PROOF
       COATING IN A COLOR OF BLACK

7. Economic analysis
    A. COSTS
       1. STEEL TUBING $2.50
       2. FASTENERS   .90
       3. RUBBER PADDING   1.40   } 5.30
       4. DRILL 4 HOLES   .20
       5. PAINTING   .30
    B. LABOR   1.10
    C. PACKAGING + CARTON   2.90
    D. PROFIT   15.70
    E. WHOLESALE PRICE   25.00
    F. RETAIL PRICE $39.00

| DESIGN ANALYSIS | NAME NO.   SECT   DATE | TIME | 11 |

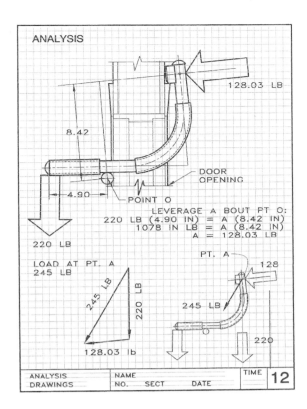

ANALYSIS

128.03 LB

8.42

DOOR OPENING

4.90

POINT O

220 LB

LEVERAGE A BOUT PT O:
220 LB (4.90 IN) = A (8.42 IN)
1078 IN LB = A (8.42 IN)
A = 128.03 LB

PT. A

LOAD AT PT. A
245 LB

128

245 LB

220

128.03 lb

| ANALYSIS DRAWINGS | NAME NO.   SECT   DATE | TIME | 12 |

Sheet 12: Graphical and analytical methods can be applied to loads and forces that must be supported by the device.

Decision

The decision table in **Sheet 13** compares two designs: the pull-up bar and the push-up grips in eight categories. Factors to be analyzed are assigned maximum values in each of these eight categories so that they will total to a maximum of 10 points. The designs are then graded in each category but not exceeding the maximum value allowed for each category. In this example, the pull-up bar earns a score of 7 compared to 5.4 for the push-up grips.

Although the table graded the pull-up bar as best, the designer still has the option of accepting this finding, rejecting it entirely, or deciding to revise portions of the designs. Human instincts should not be completely ignored during the creative process.

A design must be economically feasible and accepted by the marketplace in order for a product to be successful. Profit and loss must be considered throughout the design process, but at the final stages it becomes the primary concern. Once more, the best device for making a decision is to "put the pencil to it." This means that estimates of costs of manufacture, advertising, shipping, storage, and all other

19

DECISION

1. Decision table for evaluation
DESIGN 1: PULL—UP DEVICE
DESIGN 2: PUSH—UP GRIPS
DESIGN 3:
DESIGN 4:
DESIGN 5:

MAX	FACTORS	1	2	3	4	5
2	FUNCTION	2	1.5			
2	HUMAN FACT.	1.5	1.5			
1	MARKET ANAL	.5	.2			
1	STRENGTH	1	.5			
1	PRODUCTION	.5	.5			
1	COST	.5	.2			
2	PROFITABILITY	1.0	1.0			
0	APPEARANCE	0	0			
10	TOTALS	7	5.4			

DESIGN DECISION — NAME — NO. SECT DATE — TIME **13**

Sheet 13: Two alternative designs are evaluated in a decision table by assigning points to the important factors to be considered.

CONCLUSIONS

THE BRACKET WITH A RIB DESIGN APPEARS TO BE THE BEST AND MOST MARKETABLE SOLUTION

IT WOULD BE MOST SUITED TO THE MARKET NEEDS

GOOD POSSIBILITY OF SELLING AS ADDITIONAL PRODUCT ACCESSORY

RETAIL	$12.00
SHIPPING EXPENSES	2.00
NUMBER TO SELL TO BREAK EVEN	1000
MANUFACTURED IN HOUSE	$3.50
PROFIT PER UNIT	$2.00

RECOMMEND IMPLEMENTATION AND PRODUCTION OF THE PRODUCT AND TO LINE.

DESIGN DECISION — NAME — NO. SECT DATE — TIME **14**

Sheet 14: The final step of the of the decision process is the summary of the conclusions.

Sheet 15: The final bracket design is drawn as a working drawing ready for production.

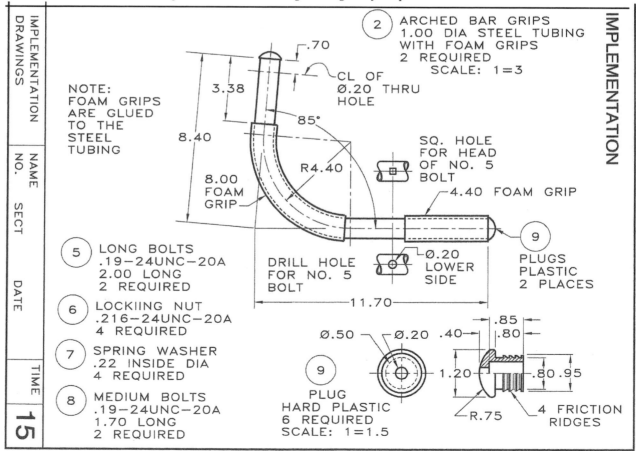

IMPLEMENTATION DRAWINGS
NAME NO. SECT DATE TIME **15**

IMPLEMENTATION

② ARCHED BAR GRIPS
1.00 DIA STEEL TUBING
WITH FOAM GRIPS
2 REQUIRED
SCALE: 1=3

NOTE:
FOAM GRIPS ARE GLUED TO THE STEEL TUBING

.70
3.38
CL OF Ø.20 THRU HOLE
85°
8.40
R4.40
8.00 FOAM GRIP
SQ. HOLE FOR HEAD OF NO. 5 BOLT
4.40 FOAM GRIP
Ø.20 LOWER SIDE
DRILL HOLE FOR NO. 5 BOLT
11.70
⑨ PLUGS PLASTIC 2 PLACES

⑤ LONG BOLTS
.19—24UNC—20A
2.00 LONG
2 REQUIRED

⑥ LOCKIING NUT
.216—24UNC—20A
4 REQUIRED

⑦ SPRING WASHER
.22 INSIDE DIA
4 REQUIRED

⑧ MEDIUM BOLTS
.19—24UNC—20A
1.70 LONG
2 REQUIRED

⑨ PLUG
HARD PLASTIC
6 REQUIRED
SCALE: 1=1.5

Ø.50 — Ø.20 .40 — .85 .80
1.20 .80 .95
R.75
4 FRICTION RIDGES

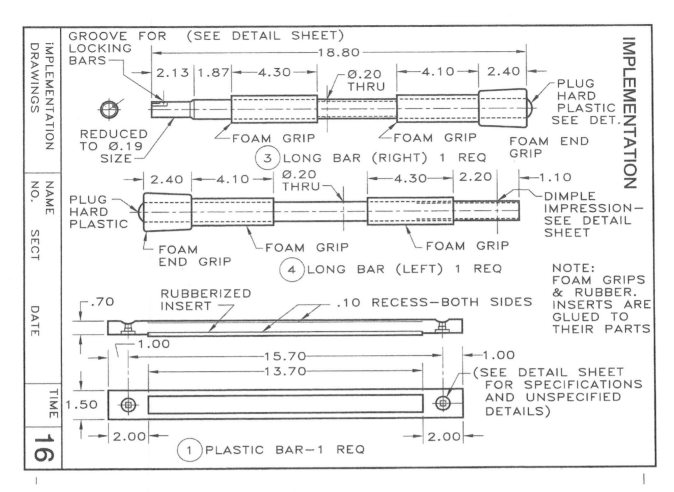

GROOVE FOR (SEE DETAIL SHEET)
LOCKING
BARS

18.80

2.13 | 1.87 |←4.30→| Ø.20
THRU

←4.10→ | 2.40

PLUG
HARD
PLASTIC
SEE DET.

REDUCED
TO Ø.19
SIZE

FOAM GRIP   FOAM GRIP   FOAM END
GRIP

③ LONG BAR (RIGHT) 1 REQ

2.40 |←4.10→| Ø.20
THRU

←4.30→| 2.20 |←1.10

PLUG
HARD
PLASTIC

DIMPLE
IMPRESSION—
SEE DETAIL
SHEET

FOAM
END GRIP   FOAM GRIP   FOAM GRIP

④ LONG BAR (LEFT) 1 REQ

NOTE:
FOAM GRIPS
& RUBBER.
INSERTS ARE
GLUED TO
THEIR PARTS

.70

RUBBERIZED
INSERT   .10 RECESS—BOTH SIDES

1.00

15.70
13.70

1.00

(SEE DETAIL SHEET
FOR SPECIFICATIONS
AND UNSPECIFIED
DETAILS)

1.50

2.00   2.00

① PLASTIC BAR—1 REQ

Sheet 16: Other parts of the product are shown in this second working drawing sheet of the implementation step of the design process. Another sheet (not shown) is needed to clarify smaller details on a separate sheet.

overhead must be listed and estimated on paper as shown briefly on **Sheet 14.** Several members of your team (if you have one) should review your estimates and give you their feedback and suggestions and share their ideas with you as well.

If the numbers look good to your team and they agree with you that the pull-up device has market appeal, then you are ready to advance to the next step: Implementation.

Implementation

The basic design of the pull-up device must be detailed as completely as possible and working drawings prepared so there will be no question as to how it will be fabricated. At this point there should be no surprises in the general design, but some minor modifications may be needed for improvement as the drawings are developed. Parts must be detailed completely in order to expedite their manufacture as shown in **Sheets 15** and **Sheet 16.**

Notes can be used to specify standard parts––the nuts, bolts, and washers–but it will be unnecessary to draw them because they will be available as standard parts. This assembly drawing **(Sheet 17)** shows how the parts fit together. Now, you need to build a prototype or model of the product and test it for function. And, let's give it a name, **Iron Gym®,** so it will have an identity of its own. **Figure 1.26** shows the final product that began as a concept and is now a reality ready for the testing and then the marketplace. Our device hooks over the upper door casing and is held in position by the leverage applied by the user's weight while doing pull-ups.

Are We There Yet?

No, there's more to do. We must test our product as shown in **Figure 1.27** and determine that it is adequately designed to serve its intended purpose, is safe to use, and explore other applicable exercises besides chinning

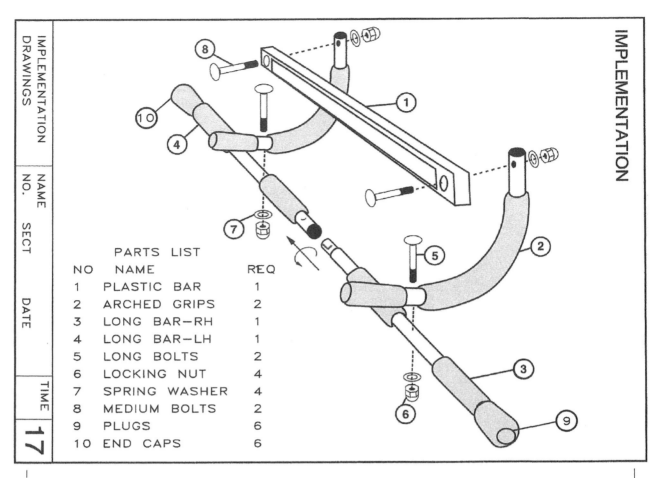

IMPLEMENTATION DRAWINGS | NAME | NO. | SECT | DATE | TIME 17

PARTS LIST

NO	NAME	REQ
1	PLASTIC BAR	1
2	ARCHED GRIPS	2
3	LONG BAR—RH	1
4	LONG BAR—LH	1
5	LONG BOLTS	2
6	LOCKING NUT	4
7	SPRING WASHER	4
8	MEDIUM BOLTS	2
9	PLUGS	6
10	END CAPS	6

Sheet 17: This partially exploded assembly drawn in perspective shows the relationship of the parts to both the fabrictor and the customer. *Ontel Products Corporation. www.getirongym.com*

that it could be used for to make it more marketable to consumers.

Alternative uses. The **Iron Gym®** lends itself to usage as an aid while doing both sit ups and push-ups. By placing the bar against the other side of a doorway, the feet will be held in position and thereby aid the exerciser while doing sit ups as shown in **Figure 1.28.** Also, the horizontal bar can be used as grips for pushups as shown in **Figure 1.29.**

Description. All products need a well-worded description that gives information and specifications regarding their features, uses, benefits, sizes, weights, color, and so forth. Look in any catalog and you will find a written description that will serve as an example. All customers want a clearly stated description of the product that they are interested in purchasing to be sure that it meets their needs and if affordable for them.

Packaging. Just about all products are packaged in a plastic bag, a cardboard box, a carton, and/or a shipping crate. Some packages are works of art designed to be appealing to the eye and attractive to the buyer. How many times have you found yourself attracted to packaging for toys at a department store? We must recognize the importance of a product's container and use it to our benefit in introducing a new product. The carton that was developed for the **Iron Gym®** is shown in **Figure 1.30** that serves as a shipping container or as an attractive box for display on a shelf in a department store with each of its six full-color surfaces filled with information about the product.

Outlets. Where will your product be sold? Do you want to be a wholesaler or a retailer? There are advantages to each. Also, many products have been sold through the Internet, by mail order, or through TV advertisements.

1.26: A photo of the finished product that has been given its own identifiable name for the marketplace. *Ontel Products Corporation. www.getirongym.com*

1.27: The primary exercise of the final design is the pull up. The device requires no permanent attachment to the door frame, it simply hooks over the upper casing on the other side. *Ontel Products Corp. www.getirongym.com*

The **Iron Gym®** lends itself very well to TV marketing where it's use can be demonstrated and then delivered by mail to its buyers.

Storage. Where are you going to put a thousand or more of these devices while you're filling orders, and how much space will it require? How long must a single unit be in storage before it is sold and how much will that cost you in the meanwhile. Never thought of that? Now is the time when you must. Maybe you should check with expense of rental space in your area to have an idea of what this will cost.

1.28: The Iron Gym® can also be used as an aid in doing sit ups in conjunction with a doorway. *Ontel Products Corporation. www.getirongym.com*

1.29: The device can also be used as handlegrips for sit ups. *Ontel Products Corp. www.getirongym.com*

1.30: The product's box can serve as a shipping box as well as a product display in a retail store. *Ontel Products Corporation. www.getirongym.com*

Office expense: Do you need a telephone, fax, correspondence stationery, desk, filing system, computer, shipping area, shipping supplies, fire insurance? And don't forget rent for your facility. All of a sudden, you're faced with the competing realities in the capitalistic system.

Employees: Someone must take the order, answer the phone, and handle the correspondence. And, who will prepare the orders for shipping one at a time or fifty at a time? Are you ready to pay Social Security for employees, contribute to their retirement plans, and cover their sick leave? What about liability insurance for your business, and what is it? Weren't we simply concerned with inventing a better mousetrap and hoping we could make a modest profit or at least break even? Welcome to the real world of capitalism, entrepreneurship, and free enterprise.

Thanks to Iron Gym®
Thanks to Justin Gelish and the folks at Ontel Products Corporation, 21 Law Drive, Fairfield, NJ 07004, Phone 973-439-07004 for allowing us to use their highly-successful **Iron Gym®** as the basis for this case study. Liberties have been taken with the application of the design process to it but, in general,` it is indicative of how it was applied to this product's development. Visit Ontel at their internet site, **www.getirongym.com,** to learn more of their products.

PROBLEMS

Most problems are to be solved on 8-1/2 x 11 inch paper, using instruments or by freehand sketching as assigned. You may use either plain paper or paper with a printed grid for laying out the problems.

Endorse each problem sheet with your name and file number, date, and problem number. Letter your responses to essay problems with single-stroke Gothic lettering as illustrated on sheets 45-47.

1. Outline your plan of activities for the weekend. Indicate aspects of your plans that you feel display creativity or imagination. Explain why.

2. Write a report not exceeding two pages on the engineer or engineering achievement that you believe exhibits a high degree of creativity. Outline the creative aspects of your choice.

3. Test your ability to recognize the need for new designs. List as many improvements as you can think of for the typical automobile. Make suggestions for implementing these improvements. Follow the same procedure for another product of your choice.

4. List as many systems as you can that affect your daily life. Separate several of these systems into their components (subsystems).

5. Subdivide the following product/systems into their individual components: (a) a classroom, (b) a wristwatch, (c) a movie theater, (d) an electric motor, (e) a coffee percolator, (f) a golf course, (g) a service station, and (h) a bridge.

6. Indicate which of the items in Problem 5 are systems and which are products. Explain your answers.

7. Make a list of products and systems that you believe would be necessary for life on the moon, in a big city, a small town, or in the woods alone.

8. You are responsible for organizing and designing a skateboard installation that will be self-supporting. Write a paragraph on each of the six steps of the design process to explain how you would apply each step to the problem. For example, explain the steps you would take to identify the problem.

9. You are responsible for designing a motorized wheelbarrow to be marketed for home use. Write a paragraph on each of the six steps of the design process, explaining how you would apply each step to the problem. For example, what action would you take to identify the problem?

10. List and explain a sequence of steps that you believe would be adequate for, yet different from, the design process steps given in this chapter. Your version of the design process may contain any of the steps discussed.

11. Design a simple device for holding a fishing pole in a fishing position while the fisherman rows the boat. Make sketches and notes to describe your design. Does such a product have a future in the marketplace? Explain.

12. Design a doorstop to keep a door from slamming into the wall behind it. Make rapid freehand sketches and notes using the six steps of the design process. Do not spend more than thirty minutes on this problem. Indicate any information you would need at the decision and implementation steps that you may not have now.

13. List factors to consider during the problem identification step for designing (a) a skillet, (b) a bicycle lock, (c) a handle for a piece of luggage, (d) a proposal for improving your grades, (e) a child's toy, (f) a stadium seat, (g) a desk lamp, (h) an umbrella, and (i) a hot dog stand.

14. Identify the problems you might encounter in developing a portable engineering travel kit to give engineers the capability of making engineering calculations, notes, sketches, and drawings. The kit might include a carrying case, calculator, computer, instruments, paper, reference material, and other accessories.

15. When sketching preliminary ideas, should you keep your developed sketches, both the good and weak ones, or keep only the good ones? What are the advantages and disadvantages of both approaches, and which position is the better one to take? Give this a little thought.

16. Make a list of problems that are in need of solution, both simple and complex. Got any ideas as to what is needed as a solution for any of them? Want to be successful? This may be your chance.

THOUGHT QUESTIONS

Suggestion: Discuss some of these topics with one or more of your classmates as a means of enriching and expanding your outlook on these general questions. Develop ideas and concepts and share them with your classmates and learn their positions in the process as well. Not all learning comes from a textbook or the classroom. Your instructor may require written responses and that would be good for you in developing your communication skills.

1. Is there a single part of the design process that is more important in obtaining a successful design than others? If so, which is it? Explain your answer.

2. Do you notice daily problems that cry out for solutions, new methods, and better designs or do you live in a perfect world? Have you often thought of solutions off the top of your head? Name several.

3. Have you ever considered modifying and improving existing designs for consideration as patentable products? Why haven't you? You must have recognized product improvements that you could make, haven't you?

4. Do you get satisfaction from being creative, inventive, and original in your work assignments? Or, had you rather perform highly structured, routine assignments with few surprises?

5. If you chose to become rich, by some means other than by winning the lottery, what paths do you see as being the best ones to take? Haven't thought about this? Why not now?

6. Give thought to where you want to be in your personal and work career at ages 30, 40, 50, and 60-plus. Do you have a plan for achievement?

7. If you are a self-starter, a hard worker with creative ideas, what do you think your prospects are going to be during your career? What aspects of your career would be most likely to turn you off and discourage you?

8. Assume that you are in charge of hiring and managing employees. Describe the qualities and traits that would be desirable in them.

9. As a personnel manager, describe the characteristics of employees that you would rank highest and most eligible for salary raises.

10. Is creativity a gift, a developed trait, or a totally unknown occurrence? There's no answer in the back of the book for this one.

11. Why is it recommended that you write down your ideas and procedures on work sheets when you are applying the design process? Can you think of a better way of progressing? If you have it, what is it?

12. Is it a waste of time to record ideas during the brainstorming process even if the ideas are not very good? Explain your answer.

13. You've used a pencil throughout your entire life; have you learned to let your ideas transfer from the point of your pencil and onto the paper? Shame on you if you haven't. Work on speed and let your ideas flow onto your work sheet. Don't spend time erasing, correcting, or checking; that will come later.

14. If you have a good idea, why waste time with the design process? Why not just design it and build it? Explain your response.

15. When our economy experiences hard times, what measures could be taken to keep one of the following businesses alive and competitive? Design office, hot dog stand, building firm, car dealership, manufacturing operation, tire shop, airline. Or, select one of your own choosing that is similar.

16. At the end this chapter the words *capitalism, entrepreneurship,* and *free enterprise* were used. You may know what these words mean, but it would be good to revisit them and incorporate them into your vocabulary. Look them up and write a paragraph to define each of them because this is where your future lies.

17. Most successful ventures require ideas, experimentation, risk taking, and creative thought, but most of all, a great deal of hard work. What is your assessment of yourself in these categories?

18. What are the advantages and disadvantages of working on a design project as a single individual as opposed to working as a member of a team with others?

19. You could have invented the Hoola Hoop, a plastic tube connected at its ends to form a circular hoop. Millions of it were sold and are still selling. It was a favorite toy among young and old alike. What factors made such a simple design so successful?

20. Childhood games, hide and seek for example, are based of the application of creativity to make them more fun and less boring. Would you rather have a career that is routine with few surprises or had you rather have a creative one? Explain.

21. What life style would you would appeal most to you? While being as realistic as possible, list the factors of an ideal career and in what jobs and career fields do these opportunities exist?

22. Do you know yourself? Do you know what your capabilities, what you're best at, what kind of career you want? Would you rather live in a city or a small town, work in large company or a small one? Describe the career that you believe to be ideal for you.

23. We are all different in varying degrees. What are the differences in attitudes and aptitudes of the people in the following career fields? Banker, engineer, inventor, carpenter, technician, salesman, and other fields of your choice.

24. Where do you want to be and be doing in the future that would make your career satisfying, productive, and challenging? Outline the pathway that will enable you to get there. Now is the best time to begin this trip.

25. Have you ever been involved in a project or a pursuit that required effort and application but was satisfying, even fun? Maybe something could be learned from that experience. What was it that caused it to be simulating. Could that be the secret of success?

The problems on this page (front and back) can be used as projects with which to apply the steps of the design process, beginning with problem identification and ending with the implementation of your solution.

Solutions to most of these examples are available in the marketplace, are being sold at a profit, and are filling a worthwhile need. Successful products provide jobs, profits, and opportunites for the designers, manufacturers, and vendors.

Most products begin as a recognition of a need for a solution to a problem, maybe a better mouse trap. From there a market might emerge leading to unforseen possibilities and opportunities.

The problems on this page, ranging from simple to more complex, should prompt you to propose a design idea of your own to develop and solve. Apply the principles and procedures covered in pages 8 through 24 and the following "Design Project" worksheets on which to record your ideas, notes, sketches, and drawings.

Remember that graphics is the medium of design without which creativity cannot be applied, developed, and transmitted to others.

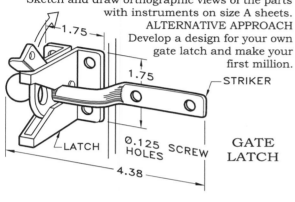

Sketch and draw orthographic views of the parts with instruments on size A sheets.

ALTERNATIVE APPROACH
Develop a design for your own gate latch and make your first million.

1.75

1.75

—STRIKER

Ø.125 SCREW HOLES

—LATCH

4.38

GATE LATCH

ELASTIC EXERCISER

Design an elastic exerciser that can be held and used as shown. Sketch your ideas on size A sheets.

ELASTIC EXERCISER

Make working drawings of your final design.

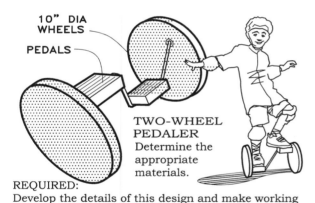

10" DIA WHEELS

PEDALS

TWO-WHEEL PEDALER
Determine the appropriate materials.

REQUIRED:
Develop the details of this design and make working drawings of it. Can you develop a design of your own?

FOLDING CLOTHES HOOK

Stainless steel
(Concept sketches)

Slotted disk

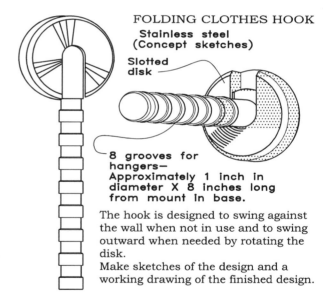

8 grooves for hangers— Approximately 1 inch in diameter X 8 inches long from mount in base.

The hook is designed to swing against the wall when not in use and to swing outward when needed by rotating the disk.

Make sketches of the design and a working drawing of the finished design.

CART ROWER

REQUIRED:
Make sketches of the parts of the rower and develop them into working drawings on size B sheets.

28 in. long
18 in. wide

Use metal for linkage components.

Use stock wheels.

Materials
Seat: Wood
Axle: Wood
Keel: Wood

BAGGAGE SIZE SCREENER

PERMITTED SIZES
Bags: 15 x 9 in.
Hanging bags:
(W+H+D)=45 in.

REQUIRED:
Design a similar device for regulating size requirements at airports. Make working drawings of your design.

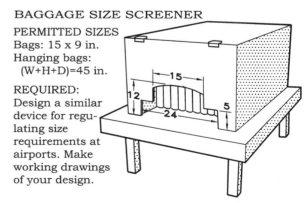

15

12

5

24

DOOR HANDLE

Design the middle stay for a warehouse door handle. Sketch orthographic views of it and the end stays and convert them into instrument drawings for fabrication on size A sheets.

Ø1.18 STEEL TUBE END STAY

3.15

1.80

1.00

MIDDLE STAY FOR HANDLES OVER 36 IN.

HOLE FOR .25 SOCKET HEAD SCREW

BUCKET STAND

Holds a 5-gal polyethylene bucket for ease of pouring.
BUCKET SIZE: Ø11.91 x Ø14.50 x 14.50 tall

Make detail drawings of the stand and determine the length of the tubing and its geometry.

Steel tubing stand: chrome plated

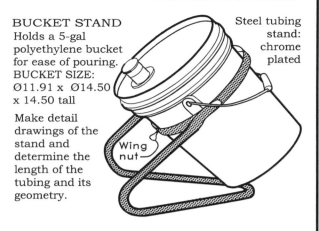

Wing nut

YARD HELPER

With a series of sketches, develop this concept into a product that can be used for general chores in the yard: hauling leaves, dirt, firewood, etc. Make working drawings of the final design.

Tubular frame

Sheet-metal body

Stock wheels

TILTING FOOTREST

Develop the details for this design, or one similar to it, and make a working drawing.
Size: 19 X 12 X 4.5-in. high.
It tilts and rocks to relieve strain at the office.

Rubber pad

Roller

Chrome-plated steel

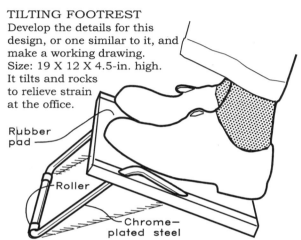

DRUM TRUCK DESIGN

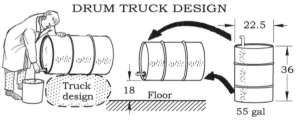

Truck design

22.5

36

18   Floor

55 gal

Design a truck for handling 55-gal drums of turpentine (7.28 lb/gal) one at a time. Design the truck to tip the drum from the vertical to the horizontal position as well as for moving the drum about the floor to different locations.

This ladder leans and sinks into soft soil, creating a hazard. Design devices that can be easily attached to it to provide more stability and safety.

Make sketches and convert them into instrument drawings on size A sheets.

Is there a market for your design?

Would you mind getting rich?

LADDER STABILIZER ATTACHMENTS

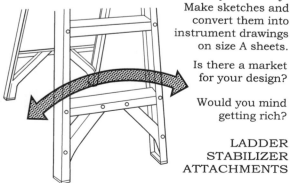

MONITOR ARM

attach to a tabletop and support a flat panel monitor (sizes 18, 19, and 21 in.). Develop your ideas with sketches and finalize your concept with instrument drawings.

How would be the price of your design?
Is there a market for it?

FLAT SCREEN

ARM

TABLE CONNECTOR

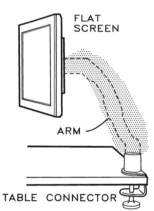

CURVY CHAISE LONGUE

Develop the details of this lightweight canvas-covered patio recliner with sketches. Determine its dimensions and make working drawings of it for production.

Canvas

Provide pillow.

Aluminum frame

Design to fold for carrying and storage.

DESIGN PROJECT

PROJECT:

COURSE: SECTION:

DATE:

FILE NUMBER:

NAME:

TEAM MEMBERS:

TITLE PAGE	NAME: FILE: SEC: DATE:	GRADE	29

PROBLEM IDENTIFICATION
1. Project title

2. Problem statement

3. Requirements

4. Needed information

5. Market considerations

PRELIMINARY IDEAS

1. Brainstorming ideas

2. Description of best ideas

3. Design sketches

Preliminary sketches

REFINEMENT

1. Description of designs

Scale drawings (continued)

ANALYSIS

1. Function

2. Human engineering

3. Market analysis

4. Physical description

5. Strength analysis

6. Production procedures

7. Economic analysis

DESIGN ANALYSIS	NAME: FILE: SEC: DATE:	GRADE	40

8. Additional analysis as needed.

DESIGN DECISION

Decision Table

Design 1:

Design 2:

Design 3:

Design 4:

Design 5:

Design 6:

Design 7:

MAX. VALUE	FACTORS TO ANALYZE	DESIGNS						
		1	2	3	4	5	6	7
	FUNCTION							
	HUMAN FACTORS							
	MARKET ANALYSIS							
	STRENGTH							
	PRODUCTION PROBS.							
	COST ANALYSIS							
	PROFITABILITY							
	APPEARANCE							
	OUTLETS/SALES							
	ADVERTISING							

10 PTS.   TOTALS:

CONCLUSIONS and RECOMMENDATIONS
(Use following sheets for the IMPLEMENTATION step.)

DESIGN DECISION

NAME:

FILE:　　　SEC:　　　DATE:

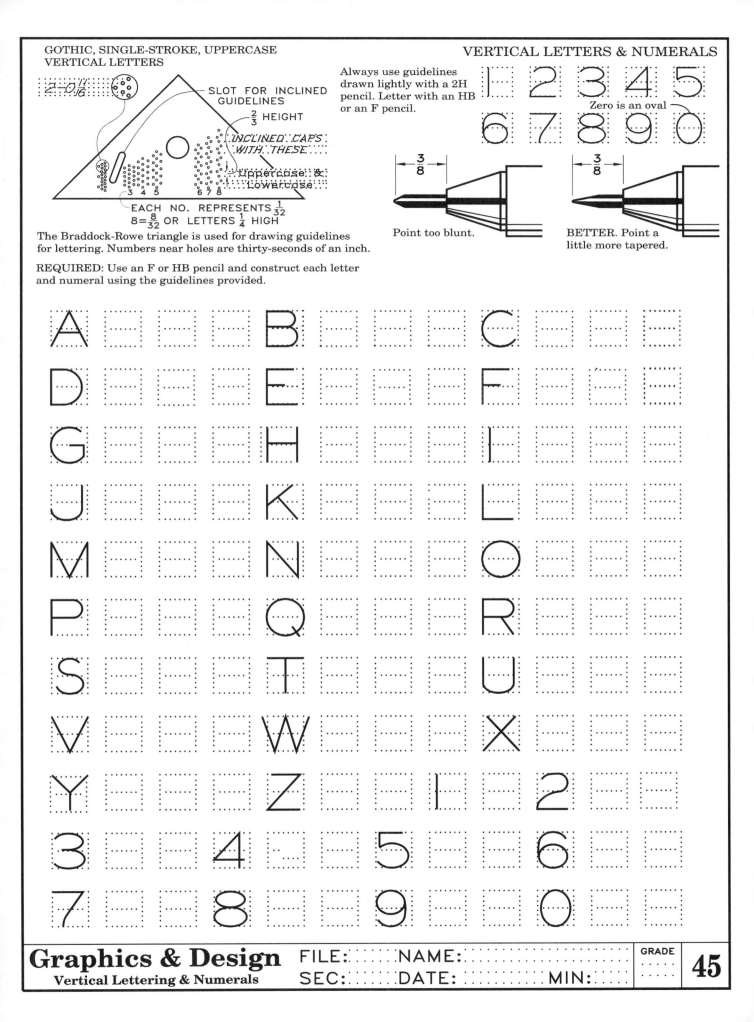

SLOT FOR INCLINED GUIDELINES

$\frac{2}{3}$ HEIGHT

INCLINED CAPS WITH THESE

Uppercase & Lowercase

EACH NO. REPRESENTS $\frac{1}{32}$
$8 = \frac{8}{32}$ OR LETTERS $\frac{1}{4}$ HIGH

The Braddock-Rowe triangle is used for drawing guidelines for lettering. Numbers near holes are thirty-seconds of an inch.

REQUIRED: Use an F or HB pencil and construct each letter and numeral using the guidelines provided.

Always use guidelines drawn lightly with a 2H pencil. Letter with an HB or an F pencil.

1 2 3 4 5

Zero is an oval

6 7 8 9 0

$\frac{3}{8}$

Point too blunt.

$\frac{3}{8}$

BETTER. Point a little more tapered.

A B C
D E F
G H I
J K L
M N O
P Q R
S T U
V W X
Y Z 1 2
3 4 5 6
7 8 9 0

A B C D E F G
H I J K L M N
O P Q R S T U
V W X Y Z &

Letter Oh is an inclined circle

Learn the proportions of each letter.

Use an F or HB pencil and construct each letter
and numeral in the guideline spaces provided.

Always use guide-
lines drawn lightly
with a 2H pencil.
Use an HB or F
pencil for lettering.

GUIDELINES   68°   INCLINED GUIDELINES

LOWERCASE $\frac{3}{5}$ HEIGHT OF UPPERCASE

$\frac{2}{3}$ LOWERCASE HEIGHT OF UPPERCASE

Upper & Lower   Uppercase & Lowercase

The Ames guide is used for
drawing lettering guidelines
by setting the dial to the number
of thirty seconds of an inch for
the desired letter height.

EACH NO. REPRESENTS $\frac{1}{32}$
$8 = \frac{8}{32}$ OR LETTERS $\frac{1}{4}$ HIGH

A   B   C

D   E   F

G   H   I

J   K   L

M   N   O

P   Q   R

S   T   U

V   W   X

Y   Z   1   2

3   4   5   6

7   8   9   0

Graphics & Design   FILE: NAME: GRADE 46
Inclined Lettering: Text and Numerals   SEC: DATE: MIN:

A. LOWERCASE VERTICAL LETTERS

a b c d e f g
h i j k l m
n o p q r s t
u v w x y z

GOTHIC LETTERING

Always use guidelines made as lightly as possible with a 2H or 4H pencil so they do not need to be erased. Inclined letters make an angle of 67.5 degrees with the horizontal. Most engineering lettering is one-eighth inch (3 mm) high.

a b p   h.
               2h
               h.

Use an F or HB pencil and construct each letter using the guidelines provided below.

LOWERCASE LETTERING
B. LOWERCASE INCLINED LETTERS

a b c d e f g
h i j k l m
n o p q r s t
u v w x y z

1. LOWERCASE VERTICAL LETTERS

a   b   c   d   e
f   g   h   i   j
k   l   m   n   o
p   q   r   s   t
u   v   w   x   y
z

2. LOWERCASE INCLINED LETTERS

a   b   c   d   e
f   g   h   i   j
k   l   m   n   o
p   q   r   s   t
u   v   w   x   y
z   &   y

A. ARCHITECTS' SCALES

BASIC FORM SCALE: $\frac{X}{X}$ =1'–0

TYPICAL SCALES — FROM END OF SCALE

SCALE: FULL SIZE (USE 16–SCALE)

SCALE: HALF SIZE (USE 16–SCALE)

SCALE: 1=1'–0   SCALE: $\frac{1}{2}$=1'–0

SCALE: 3=1'–0   SCALE: $\frac{3}{8}$=1'–0

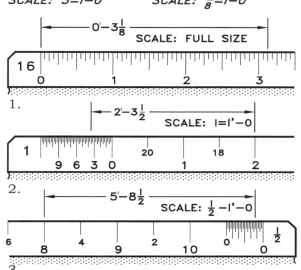

B. ENGINEERS' SCALES

BASIC FORM SCALE: 1= XX

EXAMPLE SCALES FROM END OF SCALE

10	SCALE: 1=1'	SCALE: 1=1,000
20	SCALE: 1=200'	SCALE: 1=20 LB
30	SCALE: 1=3'	SCALE: 1=3,000'
40	SCALE: 1=4'	SCALE: 1=40'
50	SCALE: 1=50'	SCALE: 1=500'

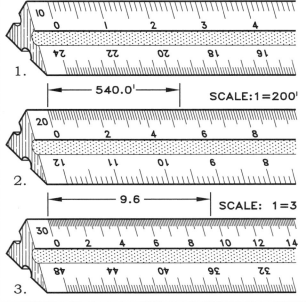

ARCHITECTS' AND ENGINEERS' SCALES

1. The partially dimensioned base flange has several of its dimensions represented by letters (A thru F). Measure the features (A thru F) and list the dimensions in the table below using the architectural scales indicated.

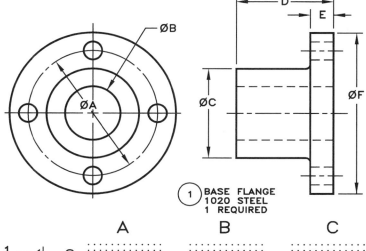

BASE FLANGE
1020 STEEL
1 REQUIRED

	A	B	C
$\frac{1}{2}$ = 1'–0			
1 = 1'–0			
$\frac{3}{4}$ = 1'–0			
$\frac{1}{8}$ = 1'–0			
FULL SIZE			

2. Following the instructions of problem 1 above, measure and list the dimensions of the features of the part and list them in the table below, using the engineers' scale as indicated.

SCALE:	D	E	F
1=1			
1=10'			
1=2			
1=30'			
1=4			
1=6			
1=0.5'			

A. METRIC SYSTEM DESIGNATION

U.S. projection where circle is visible here

European projection where circle is visible but would be hidden in U.S. projection

SI

1. Metric units and third angle of projection

SI

2. Metric units and first angle of projection

B. METRIC PREFIXES and ABBREVIATIONS

Value		Prefix	Sym.	Pronounce
$1\,000\,000 = 10^6$	=	Mega	M	"Megah"
$1\,000 = 10^3$	=	Kilo	k	"Keylow"
$100 = 10^2$	=	Hecto	h	"Heck tow"
$10 = 10^1$	=	Deka	da	"Dekah"
$1 =$				
$0.1 = 10^{-1}$	=	Deci	d	"Des sigh"
$0.01 = 10^{-2}$	=	Centi	c	"Cen'-ti"
$0.001 = 10^{-3}$	=	Milli	m	"Mill lee"
$0.000\,001 = 10^{-6}$	=	Micro	μ	"Microw"

C. METRIC SCALE FORMATS

BASIC FORM *SCALE: 1= XX*

EXAMPLE SCALES *FROM END OF SCALE*

SCALE: 1:1 (1 mm=1 mm; 1 cm=1 cm)

SCALE: 1:2 (1 mm=2 mm; 1 mm=20 mm)

SCALE: 1:3 (1 mm=30 mm; 1 mm=0.3 mm)

SCALE: 1:4 (1 mm=4 mm; 1 mm=40 mm)

SCALE: 1:5 (1 mm=5 mm; 1 mm=500 mm)

SCALE: 1:6 (1 mm=6 mm; 1 mm=60 mm)

D. METRIC MEASUREMENTS

59 mm

1 cm   SCALE: 1:1

1:1 0 1 2 3 4 5 6

1. 106 mm

1 cm   SCALE: 1:2

1:2 0 2 5 10

2. 165 mm

1 cm   SCALE: 1:3

1:3 0 3 5 10 15

3.

E. METRIC DIMENSIONING UNITS

22.0 GOOD   0.15 GOOD   .15 POOR

Missing zero

Too little room for decimal

146 GOOD   14.6 GOOD   14.6 POOR

REQUIRED: The partially dimensioned base plate has several of its dimensions represented by letters (A thru F). Measure these features (A thru F) using the engineers' scales indicated.

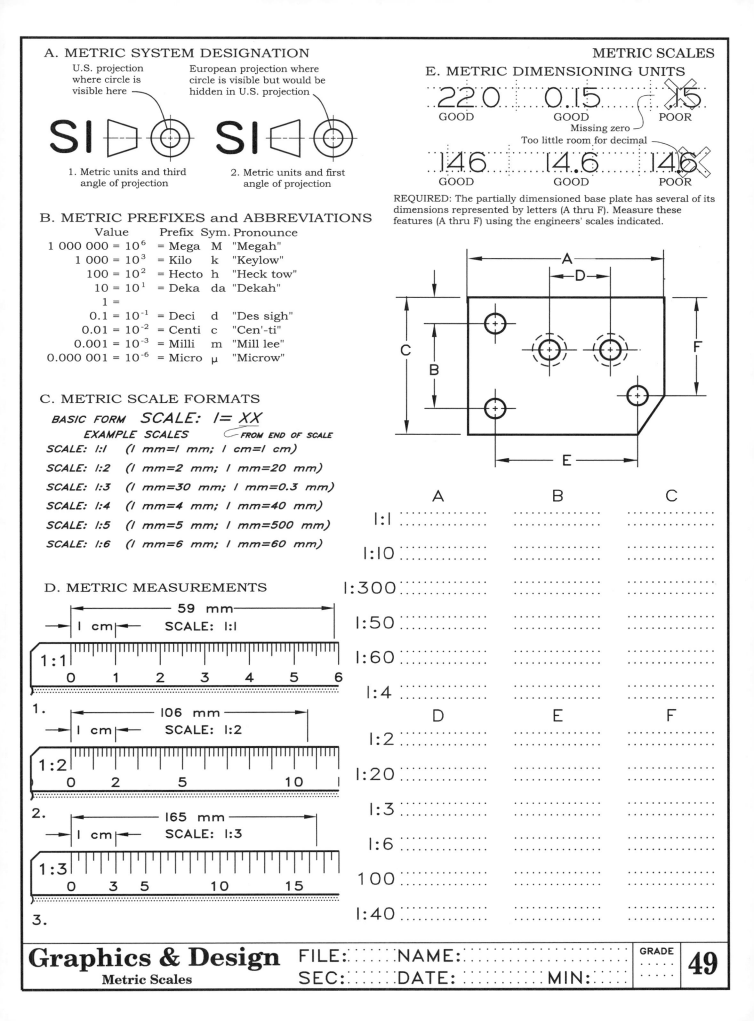

	A	B	C
1:1			
1:10			
1:300			
1:50			
1:60			
1:4			

	D	E	F
1:2			
1:20			
1:3			
1:6			
100			
1:40			

Graphics & Design
Metric Scales

FILE: NAME:

SEC: DATE: MIN:

GRADE

49

A. Architects' Scale B. Engineers' Scale C. Metric Scale

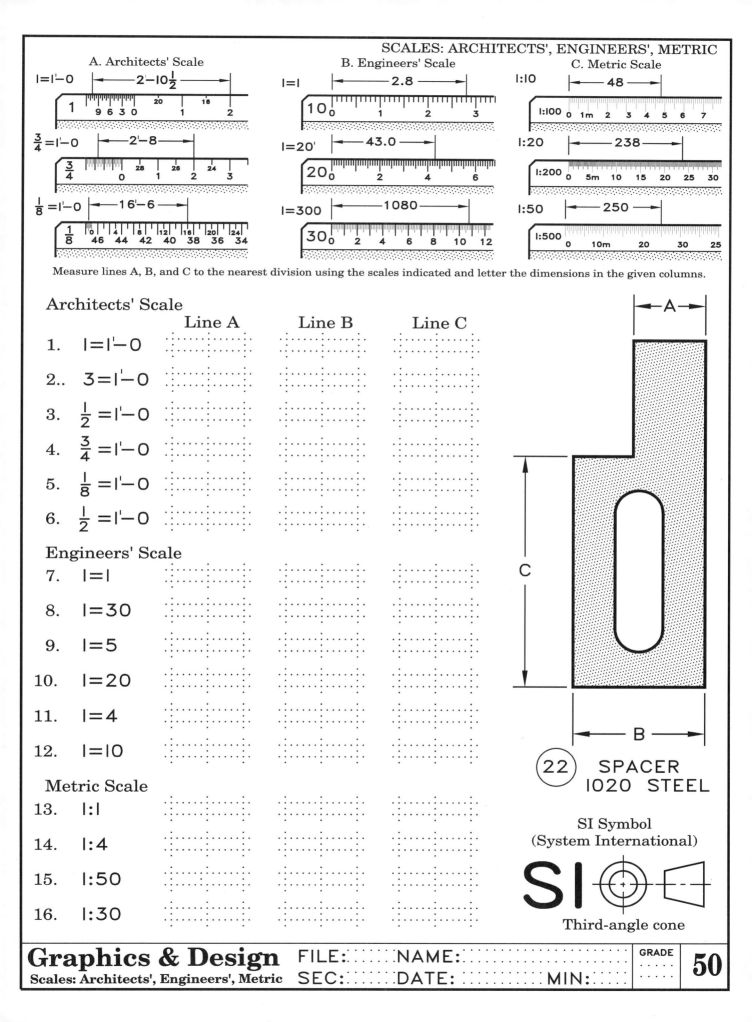

Measure lines A, B, and C to the nearest division using the scales indicated and letter the dimensions in the given columns.

Architects' Scale

		Line A	Line B	Line C
1.	$1=1'-0$			
2..	$3=1'-0$			
3.	$\frac{1}{2}=1'-0$			
4.	$\frac{3}{4}=1'-0$			
5.	$\frac{1}{8}=1'-0$			
6.	$\frac{1}{2}=1'-0$			

Engineers' Scale

7.	$1=1$			
8.	$1=30$			
9.	$1=5$			
10.	$1=20$			
11.	$1=4$			
12.	$1=10$			

Metric Scale

13.	$1:1$			
14.	$1:4$			
15.	$1:50$			
16.	$1:30$			

(22) SPACER
1020 STEEL

SI Symbol
(System International)

SI

Third-angle cone

Graphics & Design
Scales: Architects', Engineers', Metric

FILE: NAME:
SEC: DATE: MIN:

GRADE **50**

Back of the Page Problems

Draw on the back of sheet . . .

You will find this note on a number of problem sheets and sometimes you will not. Instead, your instructions may say, *"Draw four of these objects on an A-size sheet."* Sometimes your instructions will specify the page number of the sheet on which to do the drawing.

Clarification

Here's some clarification of these instructions. There are many pages that are blank, some with printed grids and others completely blank with only a printed border and a title strip for your name and other information. These are A-size (8.5" x 11") sheets that are being referred to as usable sheets for assignments.

Your instructor will probably assign the specific page by number that is best for you to use for the assignment at hand. Of course, it would be best to use the backsides of sheets that have been previously completed on the front side rather than sheets with front sides that have not yet been assigned.

Extra A-Size Sheets

Other blank sheets in addition to those in your text-problem book should be used as scratch sheets. By so doing you will preserve your drawings as future references and as examples of your work.

B-Size Sheets

Although most of your course assignments will be completed on A-size sheets, some are specified to be drawn on B-size (11" x 17") sheets. Sheets this size are provided at the rear of your book for this purpose.

Tracing-Paper Sheets

Drawings sheets on printed on semi-tansparent tracing paper are not provided as part of this book. In the recent past, drawings on tracing paper were necessary for making blue-line copies with the diazo process, but this process has been replaced by the computer.

Computer Copies of Drawings

With the advent of the computer, reproductions are made from software images for more efficiency with a higher quality than did drawings made on tracing paper. Therefore, the sheets provided for working drawings made on B-size sheets are printed on regular opaque paper rather than on translucent paper of the past.

Summary

The point of these instructions is to explain that there are many sheets on which to draw without infringing on previously solved assignments. Your instructor will specify which sheets are best suited for these assignments.

A. SELECTION OF THE PENCIL GRADE

SOFT / MEDIUM / HARD

Artistic applications

General drafting

Precise construction

PENCIL GRADE DESIGNATION

9H 8H 7H 6H 5H 4H 3H 2H H F HB B 2B 3B 4B 5B 6B 7B

B. TYPES OF PENCILS

$\frac{3}{8}$

1. Lead Holder
Holds any size lead; point must be sharpened.

2. Fine-Line Holder
Must use different-sized holders for different lead sizes; does not need to be sharpened.

3. Wood Pencil
Wood must be trimmed and lead must be pointed.

4. The Pencil Point
Sharpen to a conical point with a lead pointer or a sandpaper pad.

C. BASIC PENCIL SHARPENING

Sharpen on a sandpaper board.

Wipe the lead clean, free of graphite.

D. The Electric Eraser
is battery powered for erasing from both paper and film.

E. Standard Sheet Sizes

	Engineers'	Architects'	Metric		
A	11" X 8.5"	12" X 9"	A4	297 X 210
B	17" X 11"	18" X 12"	A3	420 X 297
C	22" X 17"	24" X 18"	A2	594 X 420
D	34" X 22"	36" X 24"	A1	841 X 594
E	44" X 34"	48" X 36"	A0	1189 X 841

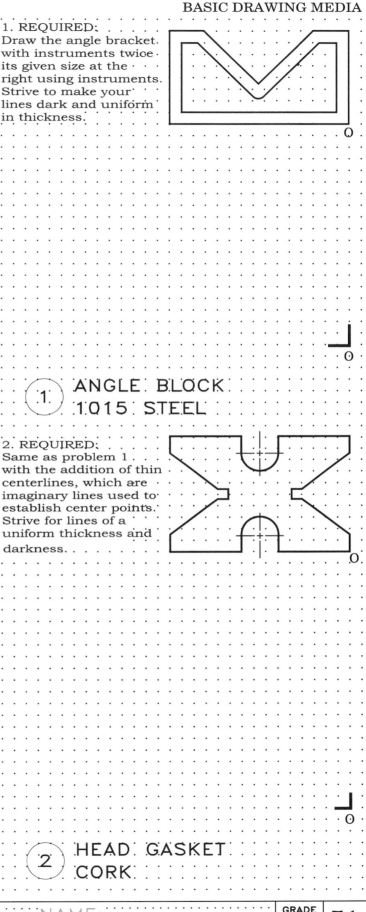

1. REQUIRED:
Draw the angle bracket with instruments twice its given size at the right using instruments. Strive to make your lines dark and uniform in thickness.

O

(1) ANGLE BLOCK 1015 STEEL

O

2. REQUIRED:
Same as problem 1 with the addition of thin centerlines, which are imaginary lines used to establish center points. Strive for lines of a uniform thickness and darkness.

O

O

(2) HEAD GASKET CORK

Computer Graphics Problems

COMPUTER GRAPHICS PROBLEMS are given on the backsides of many problem sheets of this book that can be solved and printed by several computer graphics programs. One of which is AutoCAD, an early and widely-used program.

AN EXAMPLE:
The computer drawing problem on the backside of page 52 (52B) requires that you make a double-size computer drawing of the part drawn on its grid. Space at the left of the given drawing is adequate in size for your computer-drawn version of the part drawn double size.

It would be best if you made one (or perhaps more) test prints of your drawing on 8.5 x 11 inch practice sheets before printing your final version of the problem on sheet 52B, in this case.

FILING IN YOUR NOTEBOOK.
After being graded and returned by your instructor, return the graded sheet to its appropriate place in your three-ring notebook for reference and as an example of your skill with computer graphics.

Be aware that examples of your work will reveal much about your skill to others and even your professionalism applied toward your assignments.

As when drawing by hand, draw the parts with proper line weights making visible object lines the thickest, hidden lines less thick, and center lines the thinnest. Text that is added to your drawings should be be one-eighth in height, the same as text lettered by hand. The computer makes pefection possible.

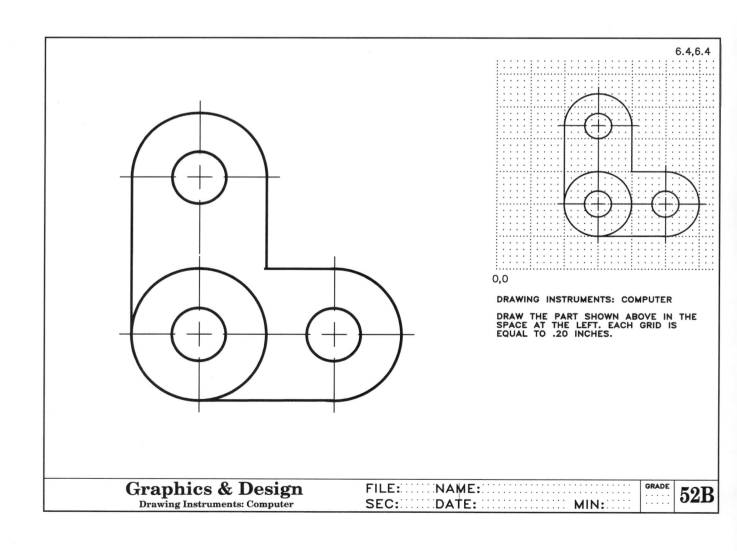

6.4,6.4

0,0

DRAWING INSTRUMENTS: COMPUTER

DRAW THE PART SHOWN ABOVE IN THE SPACE AT THE LEFT. EACH GRID IS EQUAL TO .20 INCHES.

Graphics & Design
Drawing Instruments: Computer

FILE: NAME:
SEC: DATE: MIN:

GRADE | 52B

A. THE DIVIDERS
are used to divide a line into equal segments or to transfer dimensions from one part of a drawing to another location.

PROBLEMS 1-2: Use your drawing instruments as described in the examples at the left and draw double-size views of the given orthographic views of the parts. Show centerlines for the circular holes.

1

DIVIDER PLATE

0

B. THE DIVIDERS
are helpful in transferring measurements from a scale to a location on a drawing.

+O

C. THE BOW DIVIDERS
are used for transferring measurements with greater accuracy because the settings are held by the adjusting screw.

Adjusting wheel

Threaded screw

D. THE COMPASS
is for drawing circles that are too big for the circle template. Begin by drawing the centerlines first and make several circles with slightly different radii to make the circle's lineweight the appropriate thickness.

2

UPPER LINK

0

E. LEAD SHARPENING

SHARPEN ON OUTSIDE

Sandpaper

Slight angle

Slight angle

$\frac{3}{8}$

POINT LENGTHS

BEVEL POINT

+O

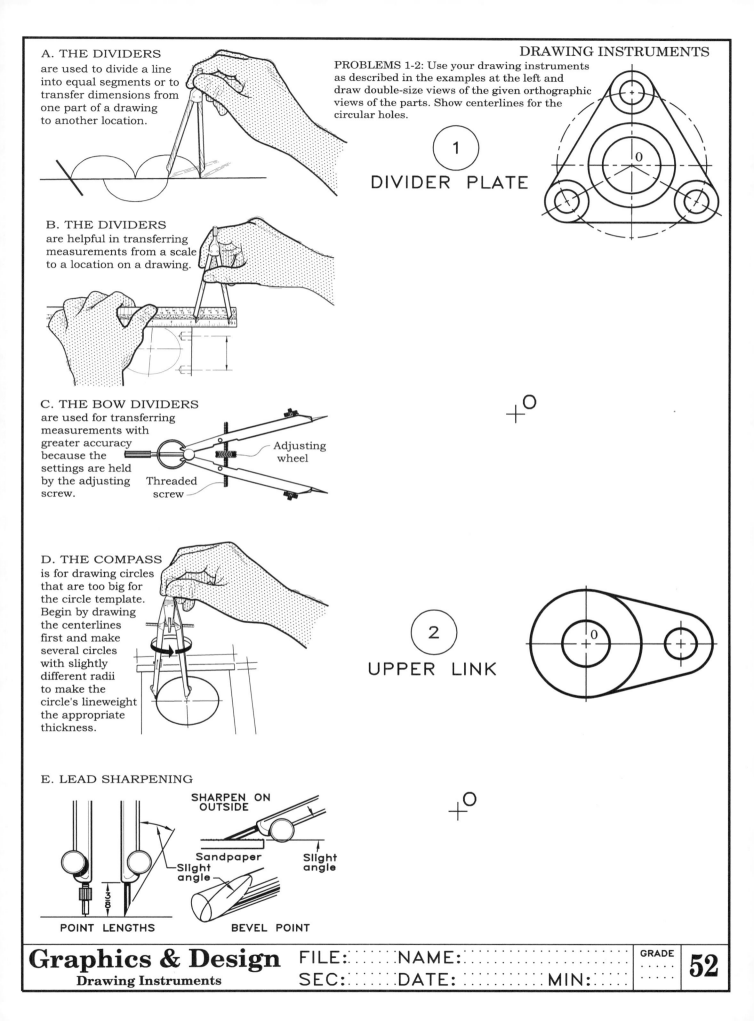

Graphics & Design
Drawing Instruments

FILE: NAME:

SEC: DATE: MIN:

GRADE

52

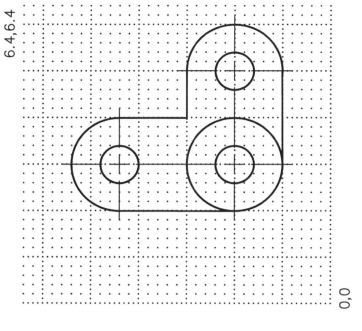

6.4,6.4

0,0

DRAWING INSTRUMENTS: COMPUTER

DRAW THE PART SHOWN ABOVE IN THE
SPACE AT THE LEFT. EACH GRID IS
EQUAL TO .20 INCHES.

Graphics & Design
Drawing Instruments: Computer

FILE: .... NAME: ....
SEC: .... DATE: ....

MIN: ....

A. TYPES OF QUADRILATERALS

4 Sides Equal	Opposite sides equal	4 Sides equal
Base (B)	Base (B)	Base (B)
1. SQUARE	2. RECTANGLE	3. RHOMBUS
Area=H(B)	Area=H(B)	Area=H(B)

Opposite sides parallel	Top (C) 2 Sides parallel	No sides parallel
Base (B)	Base (B)	
4. RHOMBOID	5. TRAPEZOID	6. TRAPEZIUM
Area=H(B)	$A = \dfrac{(B+C)H}{2}$	$A = \dfrac{(H+h)a + bH + ch}{2}$

B. INSCRIBED AND CIRCUMSCRIBED HEXAGONS

Across corners

Divide into six sectors

Hexagon inside circle

Across flats

Hexagon outside circle

Hexagon head nut

1. HEXAGON INSCRIBED

2. HEXAGON CIRCUMSCRIBED

C. GRAPHICAL DIVISION OF A LINE

A — B

Any angle

Mark off 5 divs. along AC and connect 5th mark with B

5 C

STEP 1

A 5 divisions B

Parallel

Draw lines parallel to line 5B.

5 C

STEP 2

D. APPLICATION OF GRAPHICAL LINE DIVISION

30 — Required: Divide ordinate into 3 divisions — Divide abscissa into 5 divs — 0 Abscissa — 0 5

Ordinate

STEP 1

30 Divide into 3 divisions 0 — 0 5

STEP 2

30 Divide into 5 divs 0 — 0 5

STEP 3

30 20 10 — Finished graph — 0 1 2 3 4 5

STEP 4

PROBLEM 1: Circumscribe a hexagon about the circle with its center at point A. Show construction.

PROBLEM 2: Inscribe a hexagon within the given circle with its center at point B. Show construction.

1.

+A

2.

+B

PROBLEM 3: Divide the Y-axis (ordinate) into 3 equal divisions and the X -axis (abscissa) into 5 equal divisions and label the axes as shown in the example. Plot the curve with the following data given in X and Y values: 0, 33 (given); 1, 47; 2, 50; 3, 45; 4, 34; 5, 20.

60

0

0   5

Graphics & Design
Geometric Construction

FILE: . . . . NAME: . . . .

SEC: . . . . DATE: . . . . MIN: . . . .

GRADE

53

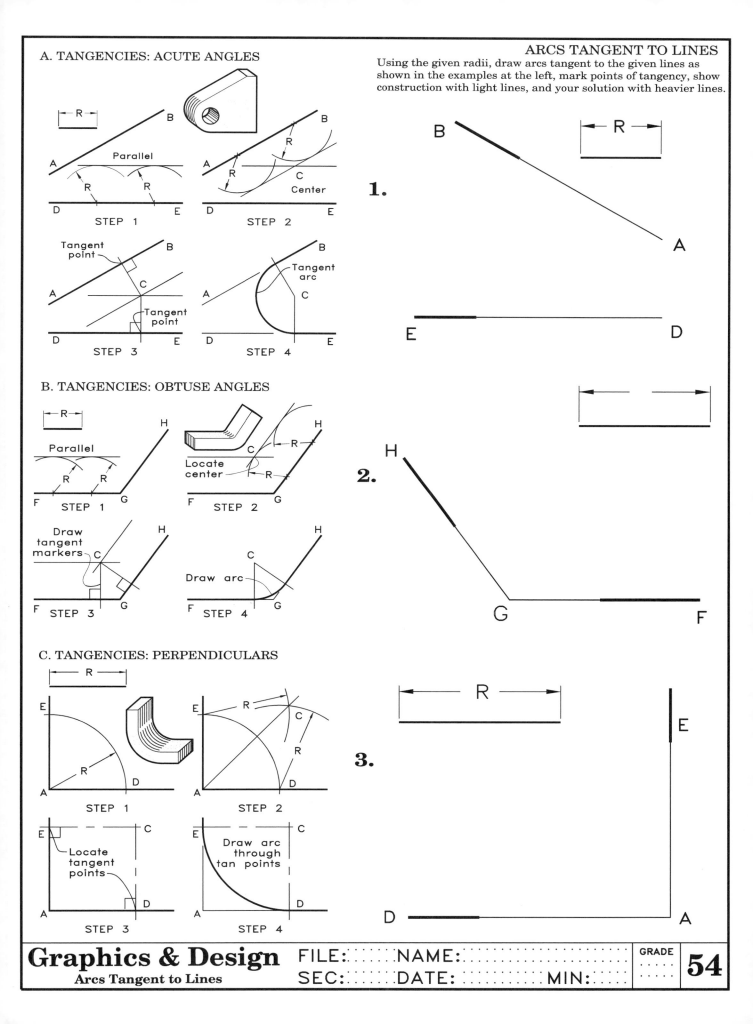

Using the given radii, draw arcs tangent to the given lines as shown in the examples at the left, mark points of tangency, show construction with light lines, and your solution with heavier lines.

A. TANGENCIES: ACUTE ANGLES

STEP 1 — Parallel, R, R, A, B, D, E

STEP 2 — R, Center, C, A, B, D, E

STEP 3 — Tangent point, Tangent point, A, B, C, D, E

STEP 4 — Tangent arc, A, B, C, D, E

B. TANGENCIES: OBTUSE ANGLES

STEP 1 — Parallel, R, R, H, F, G

STEP 2 — Locate center, R, R, C, H, F, G

STEP 3 — Draw tangent markers, C, H, F, G

STEP 4 — Draw arc, C, H, F, G

C. TANGENCIES: PERPENDICULARS

STEP 1 — E, R, D, A

STEP 2 — E, R, C, R, D, A

STEP 3 — E, Locate tangent points, C, D, A

STEP 4 — E, Draw arc through tan points, C, D, A

1.

B, A, R, E, D

2.

H, G, F

3.

R, E, D, A

Graphics & Design
Arcs Tangent to Lines

FILE: NAME:
SEC: DATE: MIN:

GRADE

54

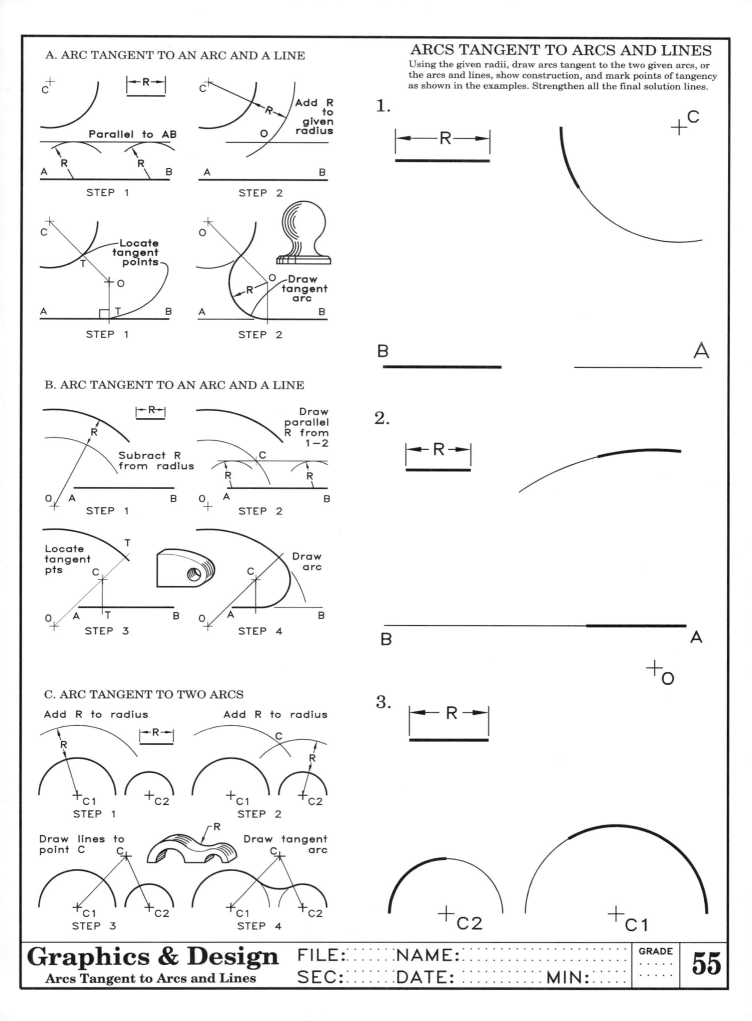

A. ARC TANGENT TO AN ARC AND A LINE

|←R→|

STEP 1 Parallel to AB

STEP 2 Add R to given radius

STEP 1 Locate tangent points

STEP 2 Draw tangent arc

B. ARC TANGENT TO AN ARC AND A LINE

|←R→|

STEP 1 Subract R from radius

STEP 2 Draw parallel R from 1-2

STEP 3 Locate tangent pts

STEP 4 Draw arc

C. ARC TANGENT TO TWO ARCS

Add R to radius | ←R→ | Add R to radius

STEP 1

STEP 2

Draw lines to point C

STEP 3

Draw tangent arc

STEP 4

1. |←R→| +C

B   A

2. |←R→|

B   A   +O

3. |←R→| +C2   +C1

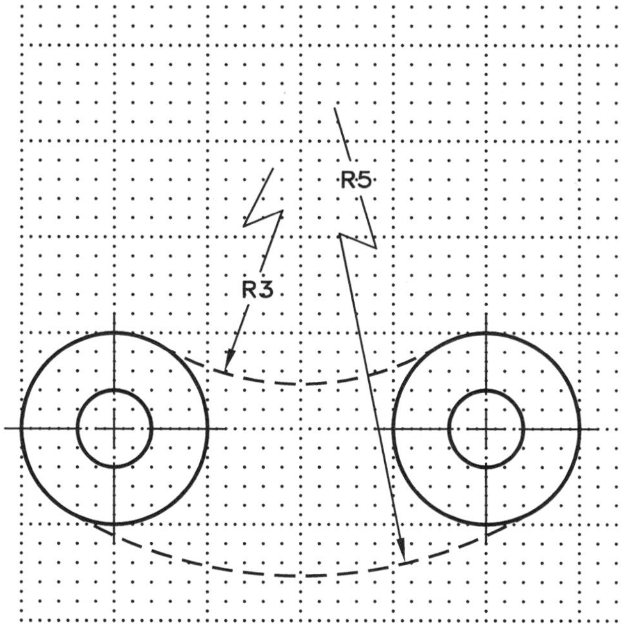

6.4,6.4

R5

R3

0,0

GEOMETRIC CONSTRUCTION

DRAW THE TWO ARCS USING THE RADII
SHOWN TO COMPLETE THE VIEW OF THE
PART. MARK THE POINTS OF TANGENGY.

Graphics & Design
Geometric Construction

FILE: . . . . NAME: . . . .

SEC: . . . . DATE: . . . .

GRADE

55B

MIN: . . . . . .

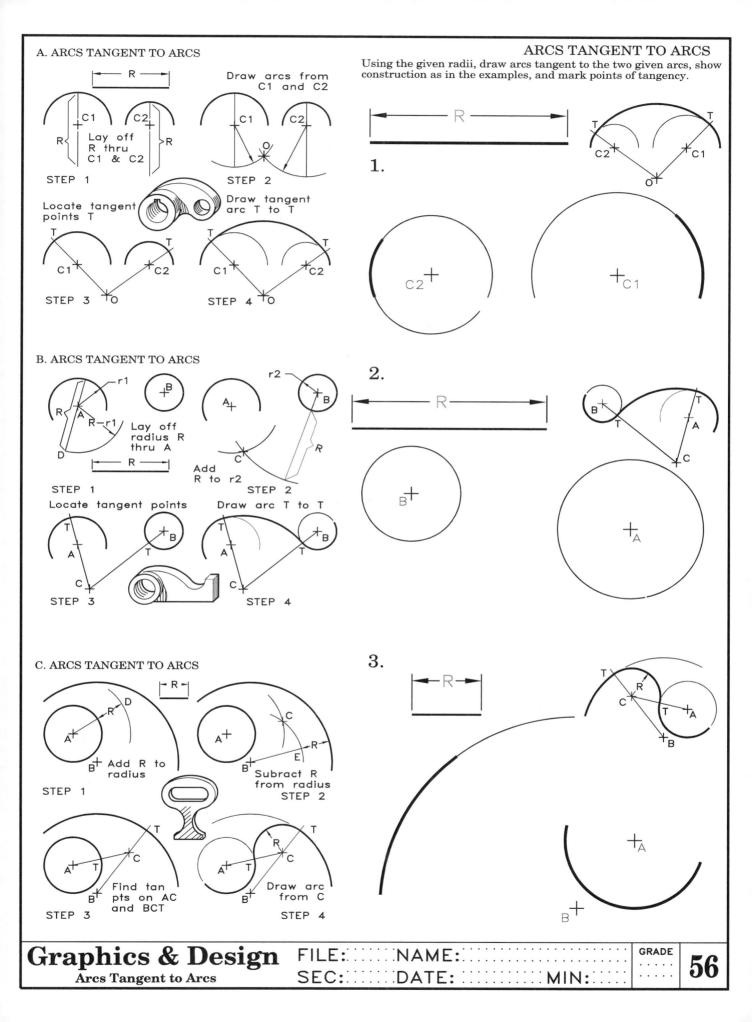

Using the given radii, draw arcs tangent to the two given arcs, show construction as in the examples, and mark points of tangency.

A. ARCS TANGENT TO ARCS

R

C1   C2

R   Lay off R thru C1 & C2   R

STEP 1

Draw arcs from C1 and C2

C1   C2

O

STEP 2

Locate tangent points T

T

C1   T   C2

STEP 3   O

Draw tangent arc T to T

C1   T   C2

STEP 4   O

1.

R

T   T

C2   C1

O

C2 +

+ C1

B. ARCS TANGENT TO ARCS

r1

R   +B

R   A   R-r1   Lay off radius R thru A

D   R

STEP 1

r2

A +   +B

Add R to r2   C   R

STEP 2

Locate tangent points

T

A

T   +B

C   T

STEP 3

Draw arc T to T

T

A   +B

C   T

STEP 4

2.

R

B +

+A

T   +A

B +   T

C

C. ARCS TANGENT TO ARCS

R

R   D

A +

B +   Add R to radius

STEP 1

C

A +

B +   E   R

Subract R from radius

STEP 2

T

A +   T   C

B +   Find tan pts on AC and BCT

STEP 3

R

A +   T   C

B +   Draw arc from C

STEP 4

3.

R

T   R

C   T   +A

B

+A

B +

1. DESIGN: FILM REEL

Use the given dimensions of the film reel and your judgement to supply those that are unspecified in the space on this sheet. Modify this design if you have a better idea.

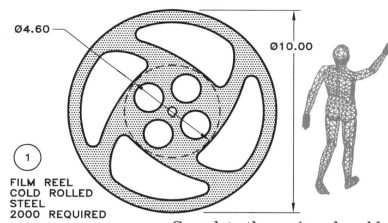

Ø4.60

Ø10.00

1

FILM REEL
COLD ROLLED
STEEL
2000 REQUIRED

2. DESIGN: HAND KNOB

Make a drawing of the descriptive view of the hand knob with instruments. You must estimate the dimensions as if you were its designer.

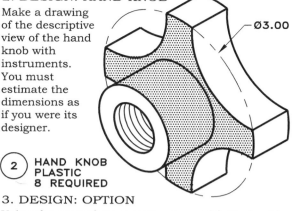

Ø3.00

2 HAND KNOB
PLASTIC
8 REQUIRED

3. DESIGN: OPTION

Using the general measurements, modify the knob's design using your own judgement and original ideas.

Complete the assigned problem in the space below.

PROBLEMS 1-7: Construct the given views on size A sheets, one problem per sheet. Select the scale that will best fit the problem to the sheet. Mark all points of tangency and strive for good line quality. (Size A sheets from previously solved problems can be used, the back of sheet 56 for example.)

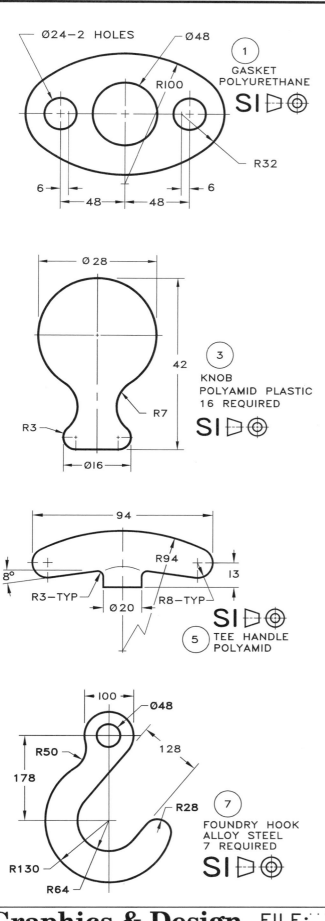

① GASKET
POLYURETHANE

Ø24–2 HOLES
Ø48
R100
R32
6 48 48 6

③ KNOB
POLYAMID PLASTIC
16 REQUIRED

Ø 28
42
R7
R3
Ø16

⑤ TEE HANDLE
POLYAMID

94
R94
13
8°
R3–TYP
Ø20
R8–TYP

⑦ FOUNDRY HOOK
ALLOY STEEL
7 REQUIRED

100
Ø48
R50
128
178
R28
R130
R64

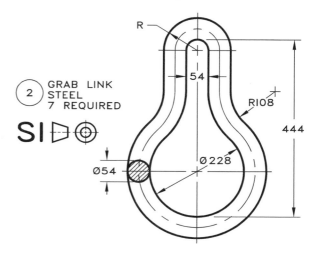

② GRAB LINK
STEEL
7 REQUIRED

R
54
R108
444
Ø54
Ø228

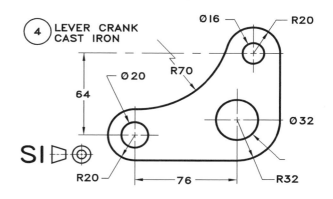

④ LEVER CRANK
CAST IRON

Ø16
R20
Ø20
R70
64
Ø32
R20
76
R32

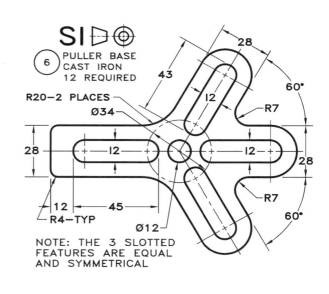

⑥ PULLER BASE
CAST IRON
12 REQUIRED

R20–2 PLACES
Ø34
43
28
60°
12
R7
28
12
12
28
R7
12
45
60°
R4–TYP
Ø12

NOTE: THE 3 SLOTTED
FEATURES ARE EQUAL
AND SYMMETRICAL

Graphics & Design
Geometric Construction: Applications

FILE: NAME:
SEC: DATE: MIN:

GRADE

58

DESIGN PROBLEMS

Approach these design problems as if you were the designer. Approximate the missing dimensions and details and draw the single descriptive view of the part assigned using true arcs of a circle with no irregular curves.

Alternative Solution: Redesign the objects with modified shapes that still conform to their major functions implied by the existing configurations. In other words, maintain the existing locations of slots, holes, and specified sizes.

DESIGN 1: EYE BOLT
Complete this partially dimensioned assembly as if you were its designer. Draw it on a size A sheet at an appropriate scale.

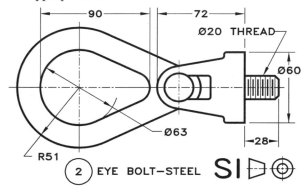

Ø20 THREAD
Ø60
Ø63
R51
28
90
72

(2) EYE BOLT—STEEL   SI

DESIGN 2: CHAIN GRAB
Make a drawing of the grab on a size A sheet using an appropriate scale. All curves are to be arcs of a circle.

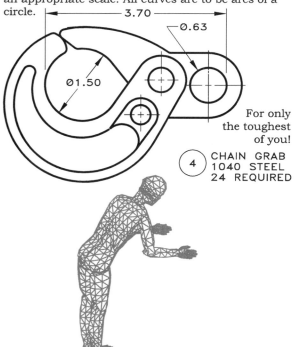

3.70
Ø.63
Ø1.50

For only the toughest of you!

(4) CHAIN GRAB
1040 STEEL
24 REQUIRED

GEOMETRIC CONSTRUCTION: DESIGN
PROBLEMS 1-6: Construct the given shapes on size A sheets, one problem per sheet. Select the scale that will best fit the problem to the sheet. Mark all points of tangency and strive for good line quality. (Draw your solution on the back of a previously used problem sheet, sheet 56 for example.)

DESIGN 3: WRENCH
Complete the design of the wrench by supplying the missing dimensions. Make all curves arcs of a circle. Draw on a size A sheet at an appropriate scale. Can you do it?

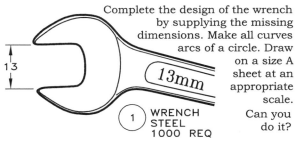

13
13mm

(1) WRENCH
STEEL
1000 REQ

DESIGN 4: LEVER
Determine the missing dimensions and draw the view of the lever at an appropriate scale on a size A sheet. Use only radii of circles. R3.00

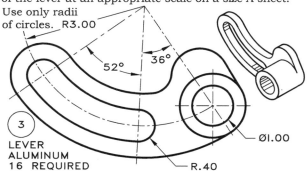

52°
36°
(3)
LEVER
ALUMINUM
16 REQUIRED
Ø1.00
R.40

DESIGN 5: 3-BOLT MOUNTING
Draw this partially-dimensioned part on a size A sheet at full size. Can you determine the radius R that is not given by applying the prinples of geometric construction?

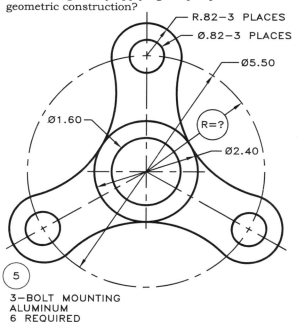

R.82—3 PLACES
Ø.82—3 PLACES
Ø5.50
Ø1.60
R=?
Ø2.40

(5)
3-BOLT MOUNTING
ALUMINUM
6 REQUIRED

Graphics & Design
Geometric Construction: Design

FILE: . . . . NAME: . . . .
SEC: . . . . DATE: . . . . . . MIN: . . . .

GRADE
. . . . .
58B

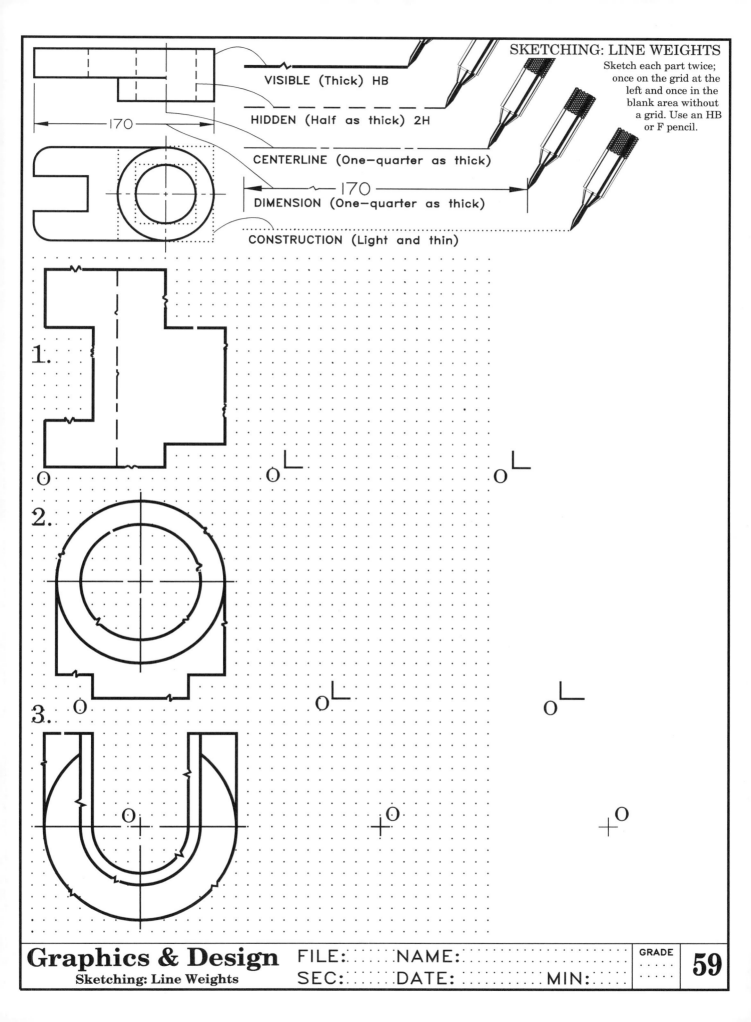

Sketch each part twice; once on the grid at the left and once in the blank area without a grid. Use an HB or F pencil.

VISIBLE (Thick) HB

HIDDEN (Half as thick) 2H

CENTERLINE (One-quarter as thick)

170

DIMENSION (One-quarter as thick)

170

CONSTRUCTION (Light and thin)

1.

2.

3.

Graphics & Design
Sketching: Line Weights

FILE:
SEC:

NAME:
DATE:

MIN:

GRADE

59

REQUIRED: Draw three views (top, front and right-side view) of each part and label the views. Count and transfer the number of grid distances to the larger grid.

If a part does not have a standard front view, use the most descriptive view as the front view. Three orthographic views are adequate to describe most parts.

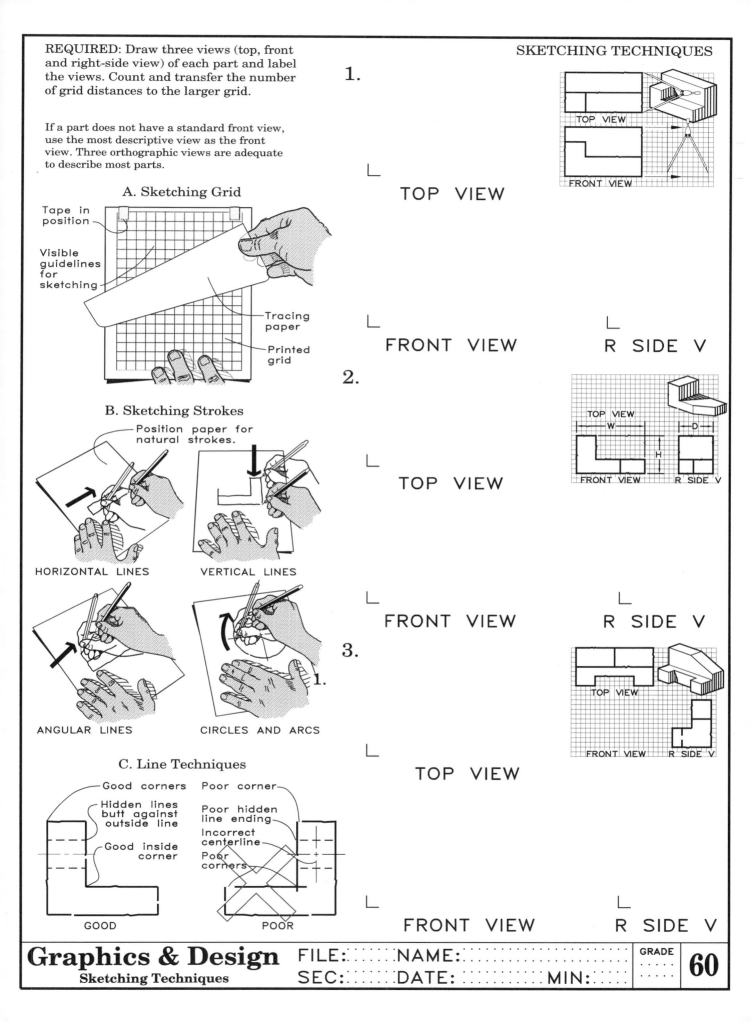

A. Sketching Grid

Tape in position

Visible guidelines for sketching

Tracing paper

Printed grid

B. Sketching Strokes

Position paper for natural strokes.

HORIZONTAL LINES

VERTICAL LINES

ANGULAR LINES

CIRCLES AND ARCS

C. Line Techniques

Good corners — Poor corner

Hidden lines butt against outside line

Poor hidden line ending

Good inside corner

Incorrect centerline

Poor corners

GOOD

POOR

1.

TOP VIEW

FRONT VIEW

R SIDE V

2.

TOP VIEW

FRONT VIEW

R SIDE V

3.

TOP VIEW

FRONT VIEW

R SIDE V

Graphics & Design
Sketching Techniques

FILE: NAME:
SEC: DATE: MIN:

GRADE

60

6.4,6.4

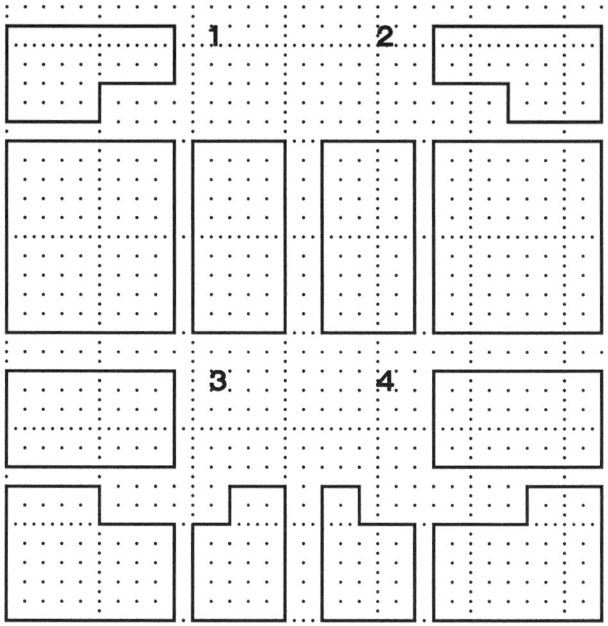

1.

2.

3.

4.

0,0

ORTHOGRAPHIC PROJECTION

PLOT THE VIEWS OF THE FOUR PARTS.
ADD THE MISSING LINES WHICH MAY BE
MISSING IN ALL VIEWS.

Graphics & Design
Orthographic Projection

FILE: NAME:
SEC: DATE: MIN:

GRADE 60B

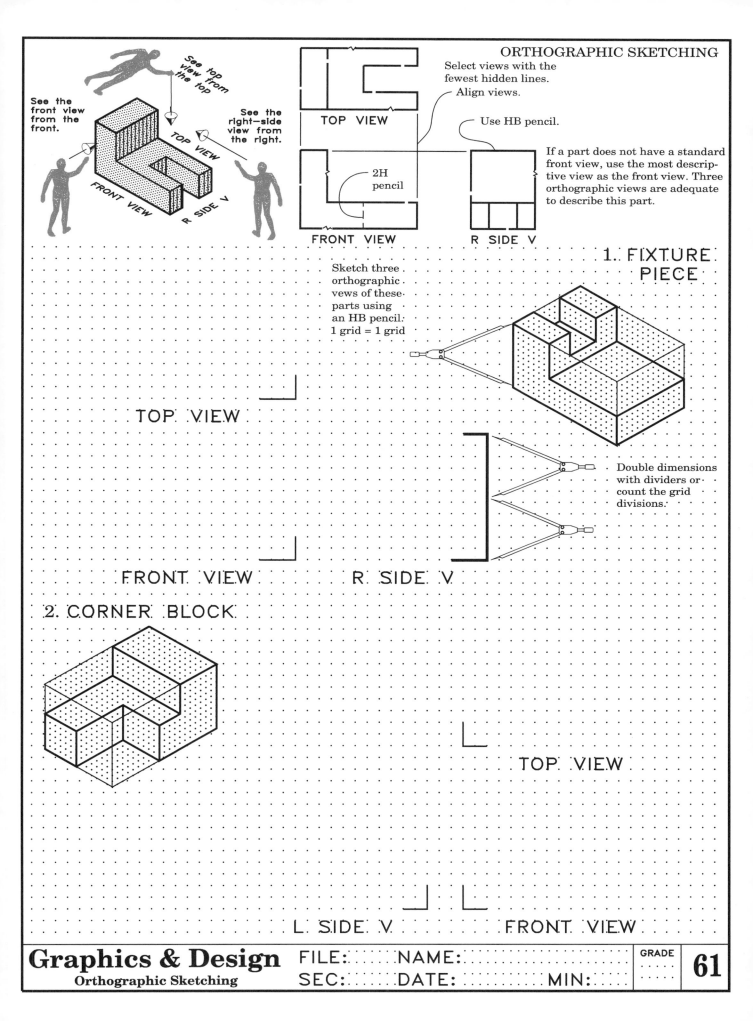

See the front view from the front.

See top view from the top

See the top view from the top

See the right-side view from the right.

TOP VIEW

FRONT VIEW

R SIDE V

TOP VIEW

FRONT VIEW

R SIDE V

Select views with the fewest hidden lines.
Align views.

Use HB pencil.

2H pencil

If a part does not have a standard front view, use the most descriptive view as the front view. Three orthographic views are adequate to describe this part.

Sketch three orthographic vews of these parts using an HB pencil. 1 grid = 1 grid

1. FIXTURE PIECE

TOP VIEW

FRONT VIEW

R SIDE V

Double dimensions with dividers or count the grid divisions.

2. CORNER BLOCK

TOP VIEW

L SIDE V.

FRONT VIEW

Graphics & Design
Orthographic Sketching

FILE: NAME:
SEC: DATE:

GRADE

MIN:

61

A. ORTHOGRAPHIC SKETCHING

This bracket can be depicted in three orthographic views as part of a working drawing from which it can be built.

REQUIRED: Sketch three orthographic of the parts and label the views as shown in the example. Strive for proper line quality.

THREE-VIEW SKETCHING

B. THE SELECTION OF VIEWS

The top, front, and side views (as shown) are selected to represent the part with the fewest hidden lines.

Loook from front to see the Front View

Look from top to see Top View

Look from right to see Right Side V

C. THE SKETCHING OF VIEWS

The views are sketched and arranged with the top view over the front view and the right-side view to the right of the front view. Height, width, and depth dimensions are placed between the views to which they apply.

A rounded inside corner is a FILLET

A rounded outside corner is a ROUND

TOP VIEW

FRONT VIEW

R SIDE VIEW

1.

TOP VIEW

FRONT VIEW   R. SIDE. V.

2.

3.

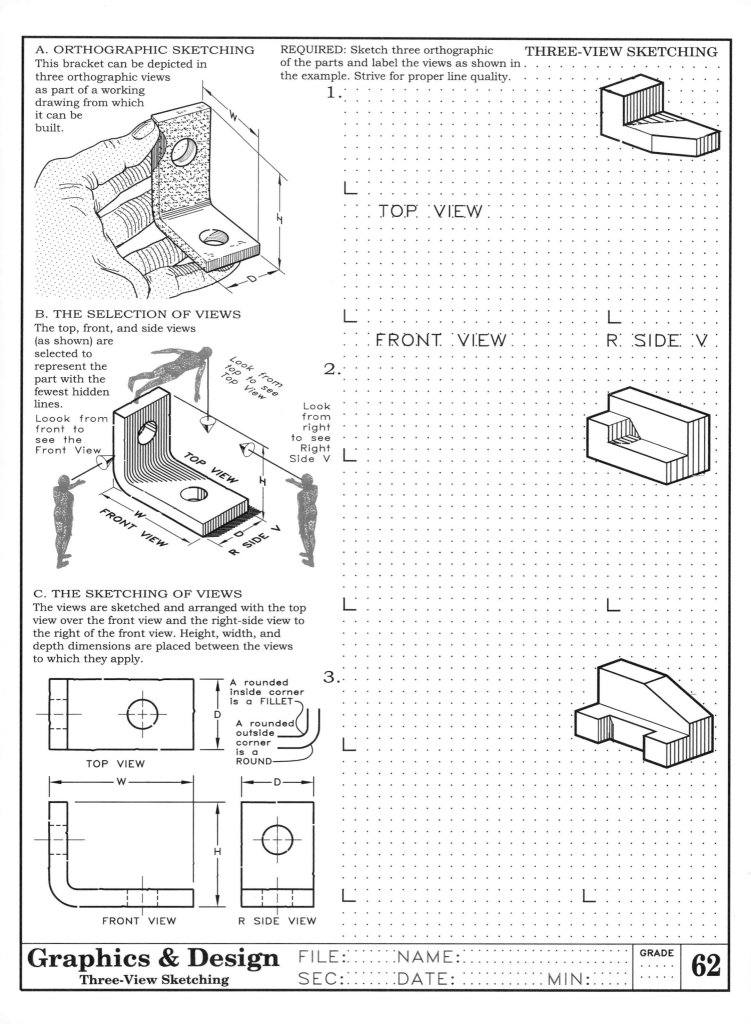

Graphics & Design
Three-View Sketching

FILE: NAME:

SEC: DATE: MIN:

GRADE

62

FILE: NAME:

SEC: DATE:

MIN:

GRADE

Sketch six orthographic views
of the assigned part. Give the
overall dimensions as letters
W, D, and H.

1 Grid = 1 Grid

TOP VIEW

TOP V

R SIDE

FRONT V

BOTTOM V

L SIDE

REAR V

REAR VIEW

TOP VIEW

RIGHT SIDE

H

D

W

LEFT SIDE

FRONT VIEW

BOTTOM VIEW

R. SIDE V

FRONT VIEW

BOTTOM VIEW

3.

L SIDE V

2.

REAR VIEW

1.

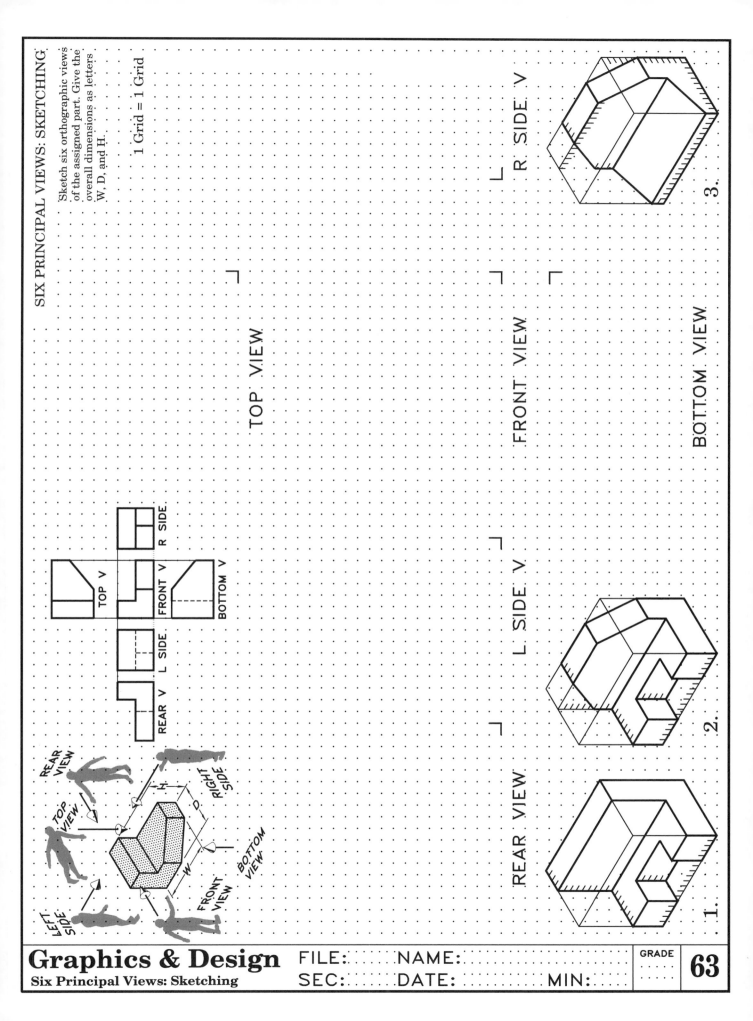

6.4, 6.4

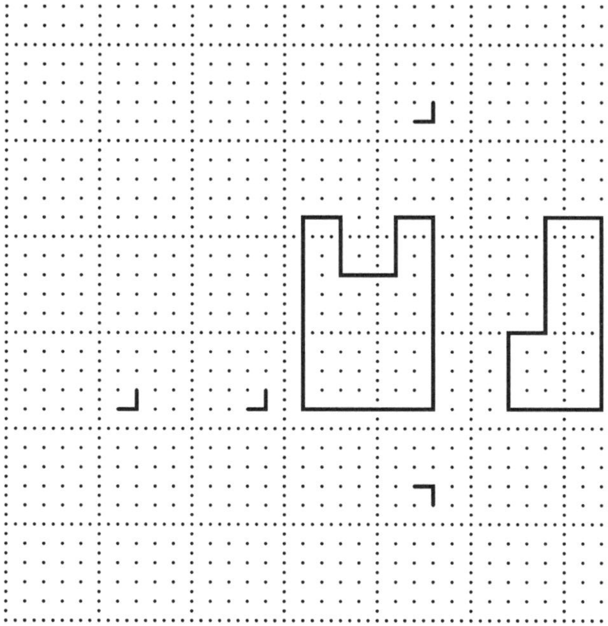

0,0

SIX-VIEW DRAWING

COMPLETE THE SIX PRINCIPAL VIEWS OF
THE PART ABOVE. LABEL THE VIEWS WITH
1/8 INCH LETTERS. USE DIMENSIONS OF
W, D, AND H THAT ARE 1/8 IN. HIGH
TO GIVE THE OVERALL DIMENSIONS.

Graphics & Design
Six-View Drawing

FILE: NAME:
SEC: DATE:

GRADE 63B
MIN:

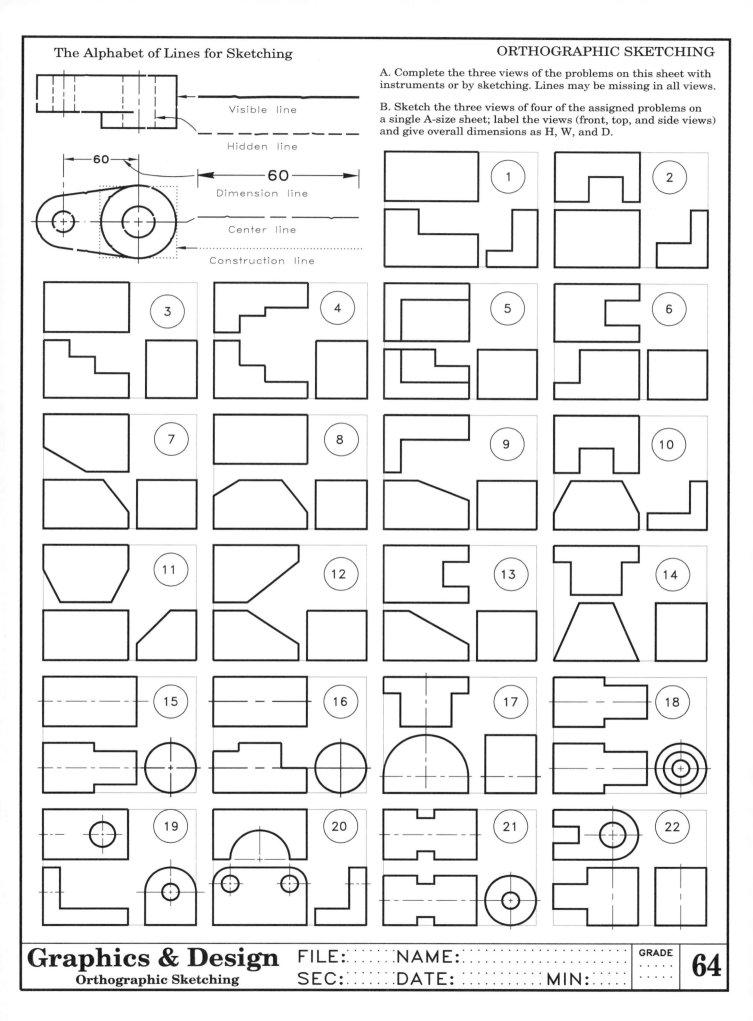

The Alphabet of Lines for Sketching

Visible line

Hidden line

60   60   Dimension line

Center line

Construction line

A. Complete the three views of the problems on this sheet with instruments or by sketching. Lines may be missing in all views.

B. Sketch the three views of four of the assigned problems on a single A-size sheet; label the views (front, top, and side views) and give overall dimensions as H, W, and D.

Graphics & Design
Orthographic Sketching

FILE: NAME:

SEC: DATE: MIN:

GRADE

64

6.4,6.4

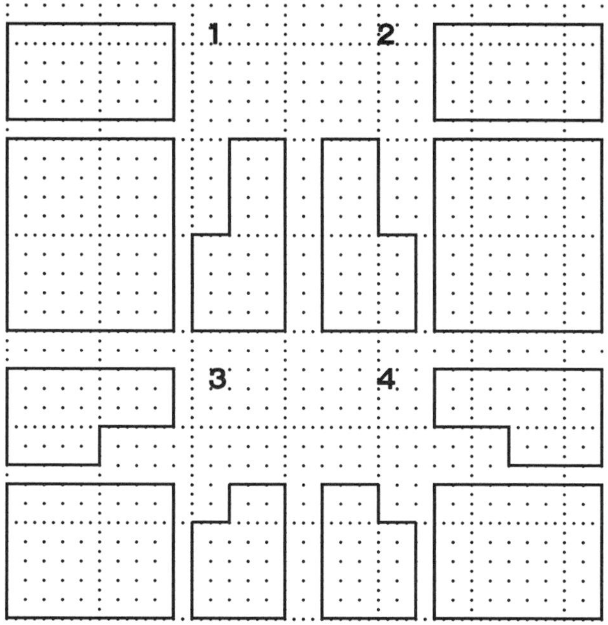

1.   2.

3.   4.

0,0

ORTHOGRAPHIC PROJECTION

PLOT THE VIEWS OF THE FOUR PARTS.
ADD THE MISSING LINES WHICH MAY BE
MISSING IN ALL VIEWS.

Graphics & Design
Orthographic Projection

FILE: ...... NAME: ......
SEC: ...... DATE: ......

GRADE **64B**

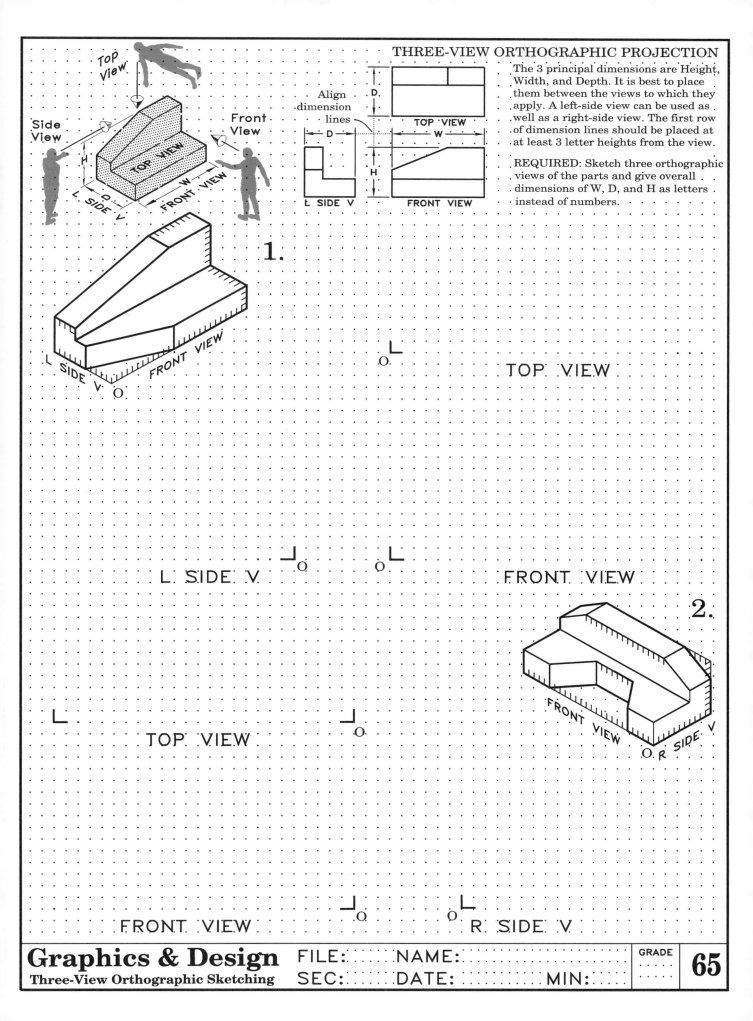

The 3 principal dimensions are Height, Width, and Depth. It is best to place them between the views to which they apply. A left-side view can be used as well as a right-side view. The first row of dimension lines should be placed at at least 3 letter heights from the view.

REQUIRED: Sketch three orthographic views of the parts and give overall dimensions of W, D, and H as letters instead of numbers.

Align dimension lines

TOP VIEW

L SIDE V

FRONT VIEW

1.

L SIDE V.   FRONT VIEW

TOP VIEW

TOP VIEW

L SIDE V.   FRONT VIEW

2.

FRONT VIEW   R SIDE V

FRONT VIEW   R SIDE V.

Graphics & Design
Three-View Orthographic Sketching

FILE: NAME:
SEC: DATE: MIN:

GRADE

65

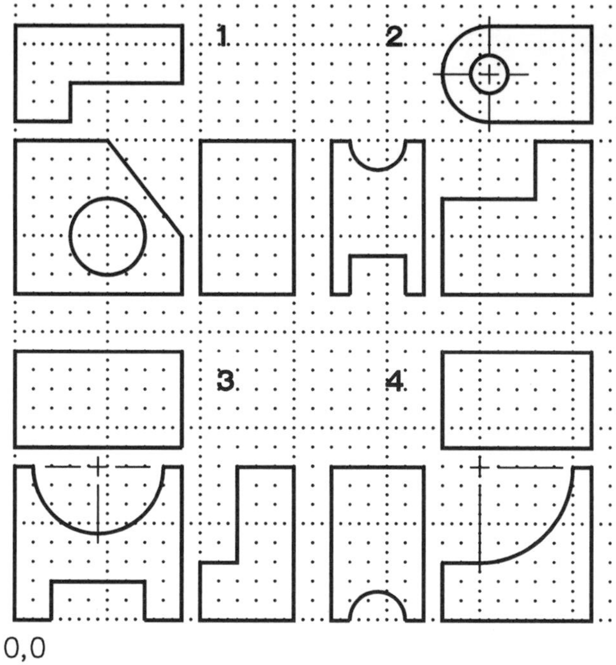

6.4,6.4

2

1

4

3

0,0

ORTHOGRAPHIC PROJECTION

PLOT THE VIEWS OF THE FOUR PARTS.
ADD THE MISSING LINES WHICH MAY BE
MISSING IN ALL VIEWS.

Graphics & Design
Orthographic Projection

FILE:
SEC:

NAME:
DATE:

GRADE

65B

MIN:

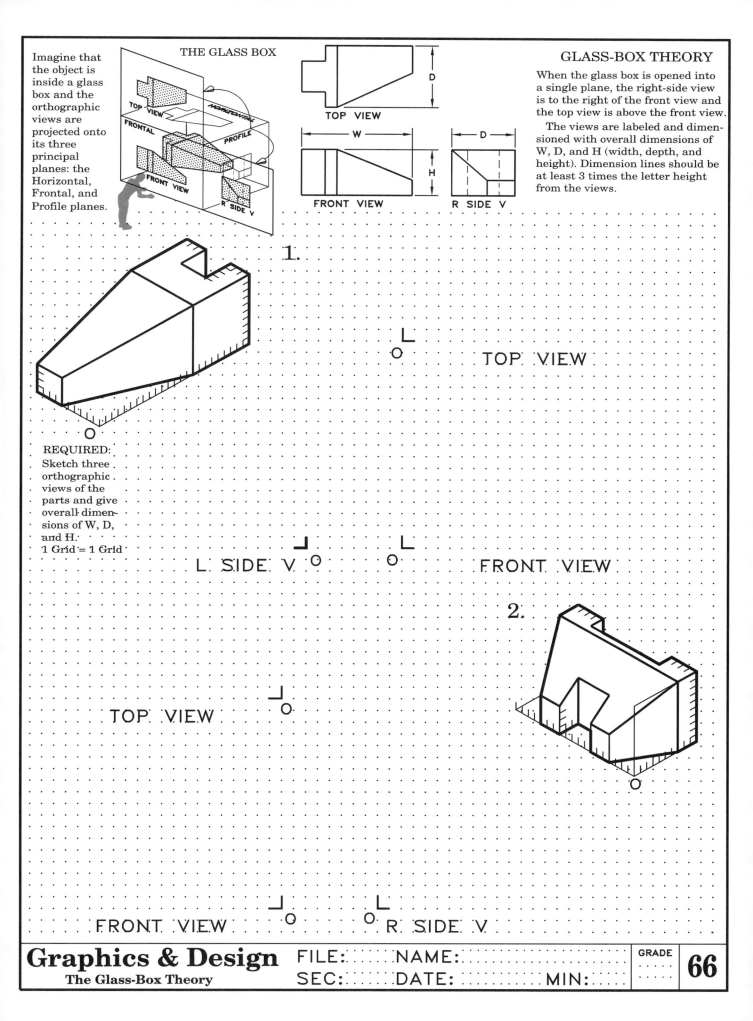

Imagine that the object is inside a glass box and the orthographic views are projected onto its three principal planes: the Horizontal, Frontal, and Profile planes.

THE GLASS BOX

GLASS-BOX THEORY

When the glass box is opened into a single plane, the right-side view is to the right of the front view and the top view is above the front view.

The views are labeled and dimensioned with overall dimensions of W, D, and H (width, depth, and height). Dimension lines should be at least 3 times the letter height from the views.

TOP VIEW

FRONT VIEW

R SIDE V

1.

TOP VIEW

REQUIRED:
Sketch three orthographic views of the parts and give overall dimensions of W, D, and H.
1 Grid = 1 Grid

L. SIDE V.

FRONT VIEW

2.

TOP VIEW

FRONT VIEW

R. SIDE V.

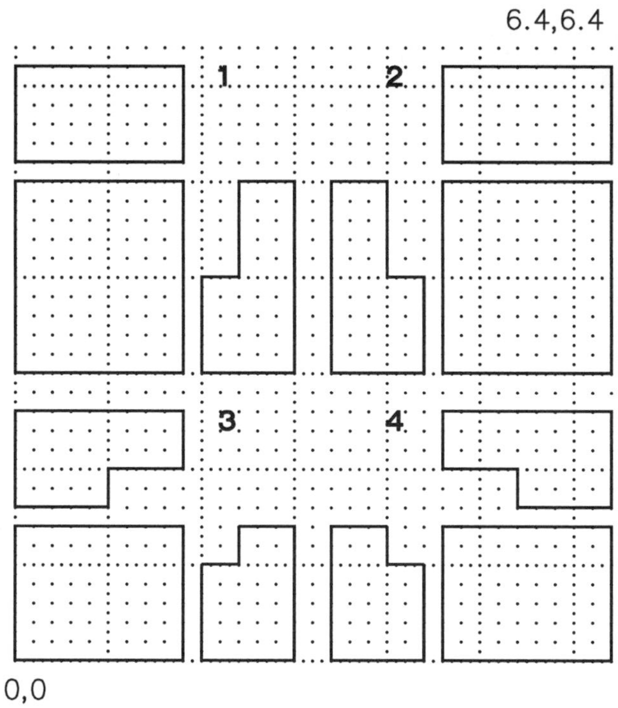

6.4,6.4

2

1

4

3

0,0

ORTHOGRAPHIC PROJECTION

PLOT THE VIEWS OF THE FOUR PARTS.
ADD THE MISSING LINES WHICH MAY BE
MISSING IN ALL VIEWS.

Graphics & Design
Orthographic Views

FILE: .... NAME: ....
SEC: .... DATE: ....

GRADE **66B**

MIN: ....

7

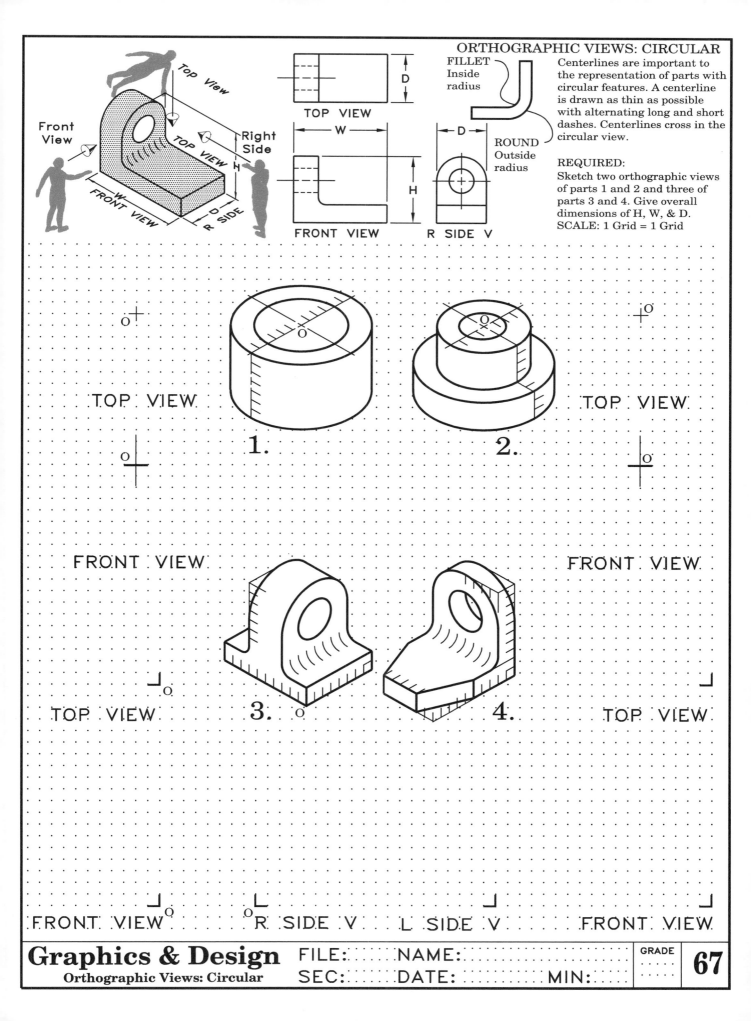

Top View

Front View

Right Side

TOP VIEW

D

TOP VIEW

W

FRONT VIEW

FILLET
Inside
radius

ROUND
Outside
radius

D

H

R SIDE V

Centerlines are important to the representation of parts with circular features. A centerline is drawn as thin as possible with alternating long and short dashes. Centerlines cross in the circular view.

REQUIRED:
Sketch two orthographic views of parts 1 and 2 and three of parts 3 and 4. Give overall dimensions of H, W, & D.
SCALE: 1 Grid = 1 Grid

TOP VIEW   1.

TOP VIEW   2.

FRONT VIEW

FRONT VIEW

TOP VIEW   3.

4.   TOP VIEW

FRONT VIEW   R SIDE V   L SIDE V   FRONT VIEW

Graphics & Design
Orthographic Views: Circular

FILE: NAME:
SEC: DATE: MIN:

GRADE

67

6.4,6.4

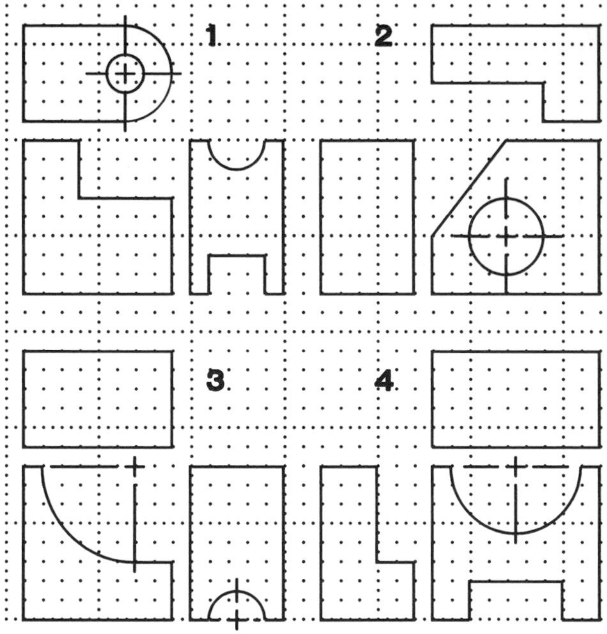

0,0

ORTHOGRAPHIC PROJECTION

**PLOT THE VIEWS OF THE FOUR PARTS.
ADD THE MISSING LINES WHICH MAY BE
MISSING IN ALL VIEWS.**

Graphics & Design
Orthographic Projection

FILE: .... NAME: ....
SEC: .... DATE: ....

GRADE 67B

BRACKET SUPPORT

Overall dimensions are given in millimeters on the two views of the bracket support. Make a 3-view orthographic drawing of the part to fit within the space below. Omit dimensions and notes. The two holes are 20 mm in diameter and the large hole is 30 mm in diameter. Either sketch or draw your solution with instruments as assigned.

SCALE: Each grid represents 4 mm.

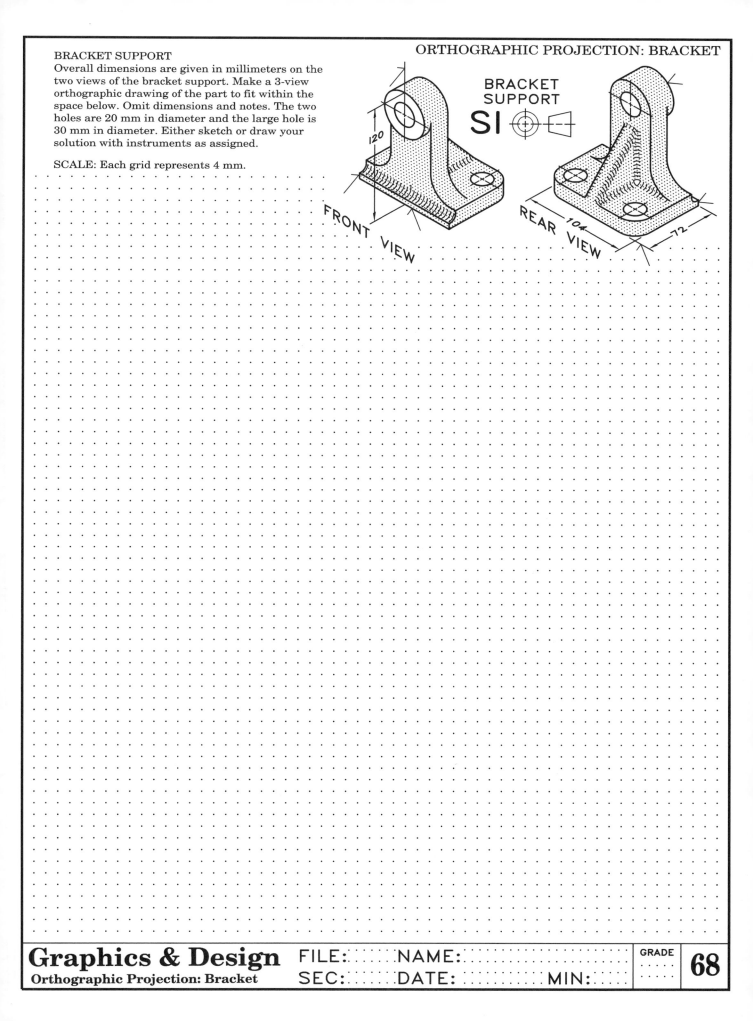

BRACKET
SUPPORT
SI

FRONT VIEW

120

REAR VIEW

104

72

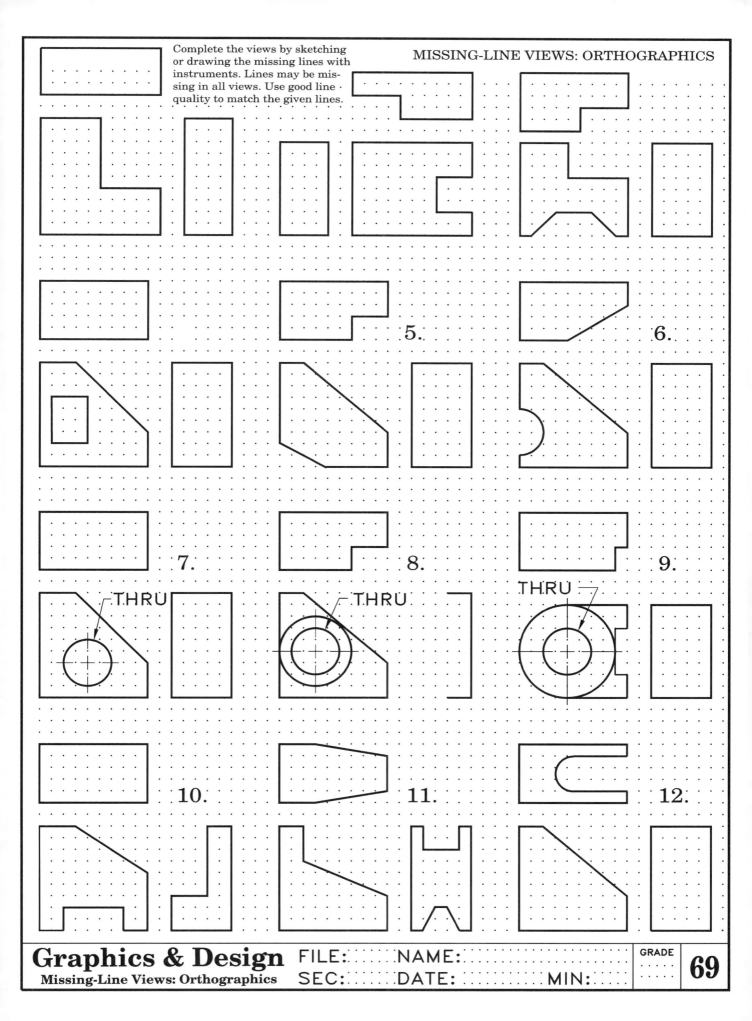

Complete the views by sketching or drawing the missing lines with instruments. Lines may be missing in all views. Use good line quality to match the given lines.

MISSING-LINE VIEWS: ORTHOGRAPHICS

5.

6.

7.

8.

9.

THRU

THRU

THRU

10.

11.

12.

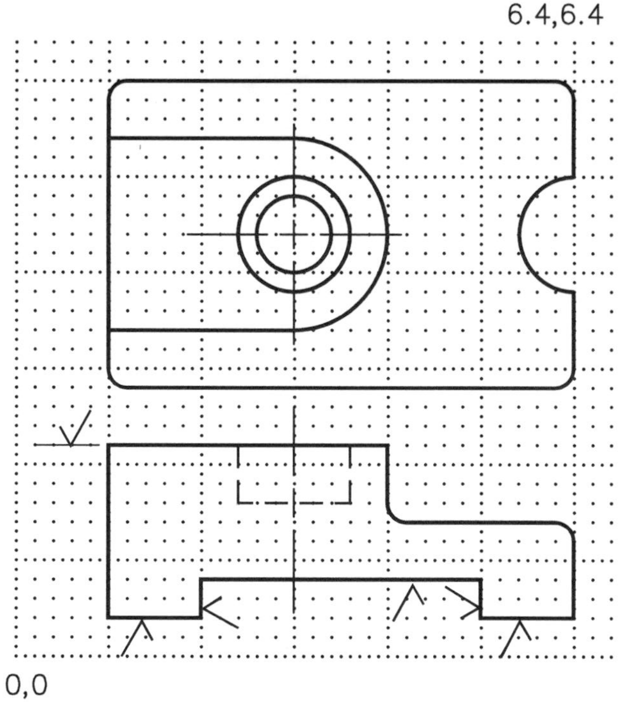

6.4, 6.4

0,0

ORTHOGRAPHIC PROJECTION

NOTE:
ADD THE MISSING LINES WHICH MAY BE
MISSING IN BOTH VIEWS.

THE "V" MARKS THAT ARE APPLIED TO
EDGES OF THE SURFACES ARE FINISH
MARKS, INDICATING THOSE THAT ARE TO
MACHINED TO SMOOTH SURFACES.

Graphics & Design
Orthographic Views

FILE: NAME:
SEC: DATE: MIN:

GRADE 69B

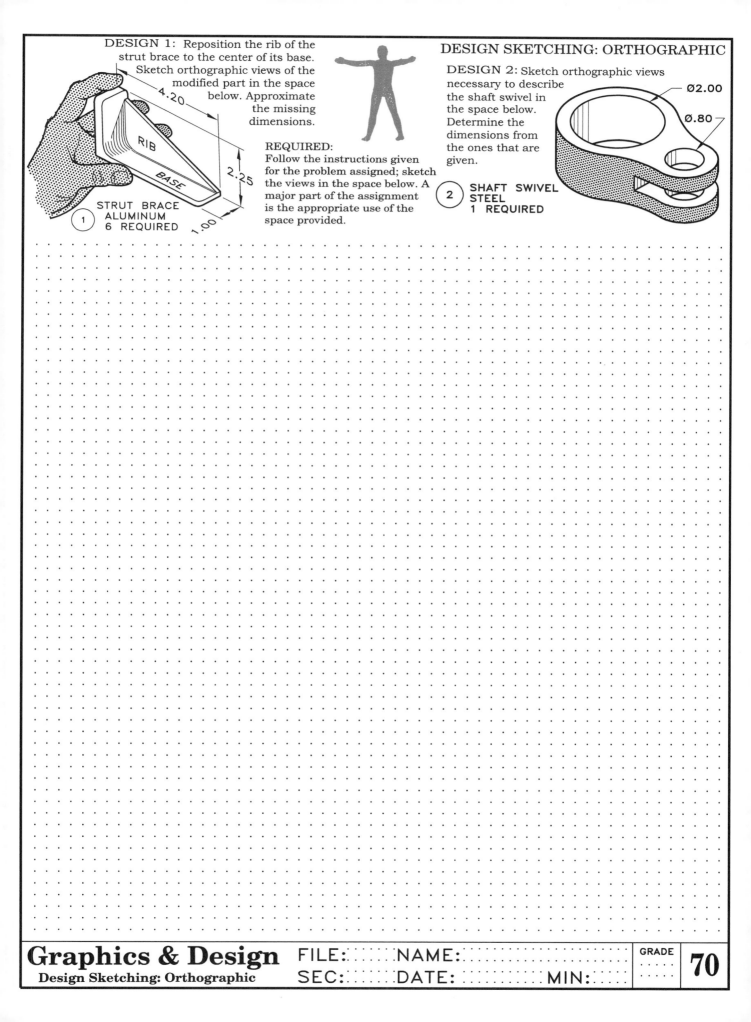

DESIGN 1: Reposition the rib of the strut brace to the center of its base. Sketch orthographic views of the modified part in the space below. Approximate the missing dimensions.

4.20

RIB

BASE

2.25

1.00

STRUT BRACE
ALUMINUM
6 REQUIRED

1

REQUIRED:
Follow the instructions given for the problem assigned; sketch the views in the space below. A major part of the assignment is the appropriate use of the space provided.

DESIGN SKETCHING: ORTHOGRAPHIC

DESIGN 2: Sketch orthographic views necessary to describe the shaft swivel in the space below. Determine the dimensions from the ones that are given.

Ø2.00

Ø.80

2

SHAFT SWIVEL
STEEL
1 REQUIRED

Graphics & Design
Design Sketching: Orthographic

FILE:
SEC:

NAME:
DATE:

MIN:

GRADE

70

6.4,6.4

0,0

ORTHOGRAPHIC PROJECTION

PLOT THE TWO VIEWS OF THE PART.
ADD THE MISSING LINES WHICH MAY
BE MISSING IN BOTH VIEWS.

Graphics & Design
Orthographic Projection

FILE: .... NAME: ....

SEC: .... DATE: ....

GRADE 70B

MIN: ....

A. ISOMETRICS: SLOPING PLANES

GIVEN

STEP 1

STEP 2

STEP 3

B. ISOMETRICS: TWO SLOPING PLANES

GIVEN

STEP 1

STEP 2

STEP 3

3. ISOMETRICS: CYLINDRICAL PARTS

STEP 1: Block in and sketch top ellipse.

STEP 1: Sketch bottom ellipse.

STEP 1: Con— nect the two ellipses.

REQUIRED:

1. Adjuster and Slider: Sketch isometric views of both the adjuster and slider that fill the available gridden space on this sheet.

2. Oblique Pictorial (alternative): Make freehand oblique pictorial sketches of the slider and the adjuster on a separate A-size sheet.

1. Adjuster

FRONT VIEW

R SIDE V

2. Slider

FRONT VIEW

R SIDE V

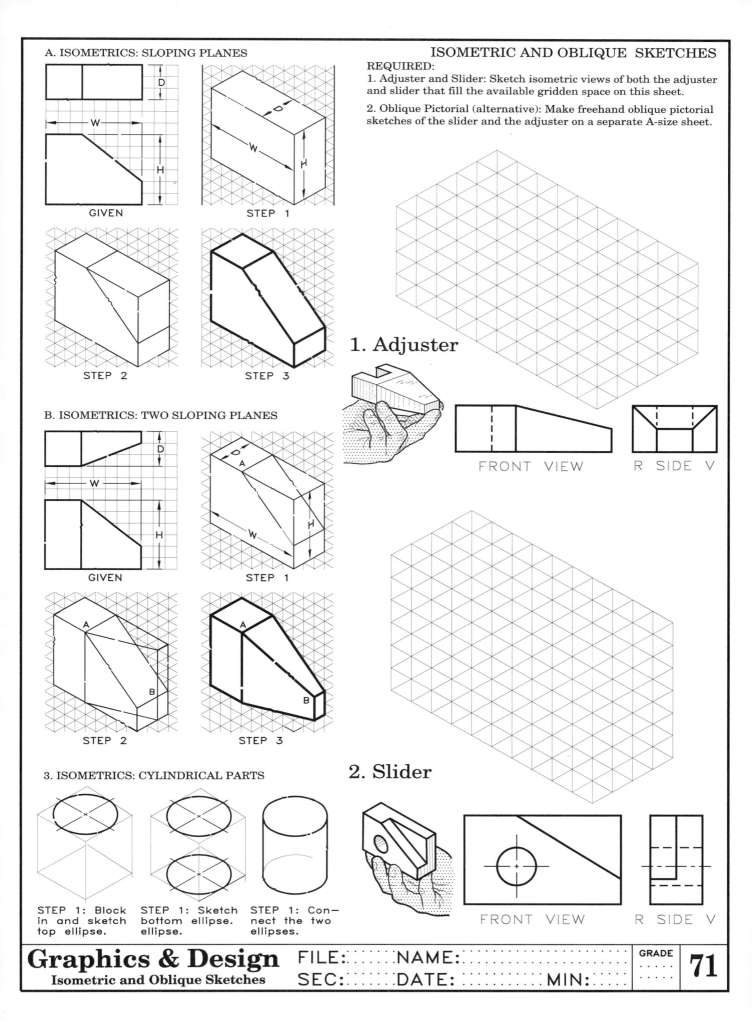

Graphics & Design
Isometric and Oblique Sketches

FILE: NAME:

SEC: DATE: MIN:

GRADE **71**

REQUIRED:
1. Base Hitch: Make a freehand sketch of the base hitch at a larger scale using the isometric guidelines.

2. Shaft Guide: Make a freehand isometric sketch of the shaft guide at a larger scale using the isometric isometric guidelines given.

1. Base Hitch

A. Sketching Circular Features

D

W

H

GIVEN

Light lines

STEP 1

STEP 2

Darken lines

STEP 3

Sketch preliminary lines lightly, sufficiently light for erasing to be unneeded, and draw final lines with an F or HB pencil.

2. Shaft Guide

B. Circular Features in Isometric

Center-lines

Tangent

STEP 1   STEP 2   STEP 3

1. CIRCLES ON A FRONTAL PLANE

STEP 1   STEP 2   STEP 3

2. CIRCLES ON A HORIZONTAL PLANE

STEP 1   STEP 2   STEP 3

3. CIRCLES ON A PROFILE PLANE

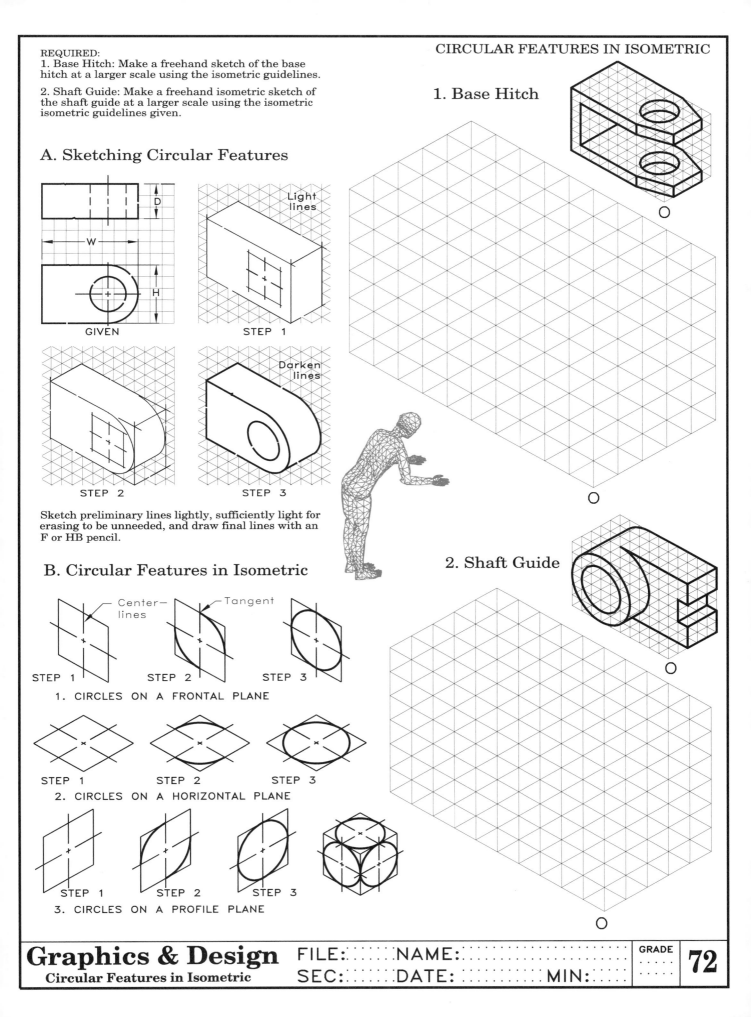

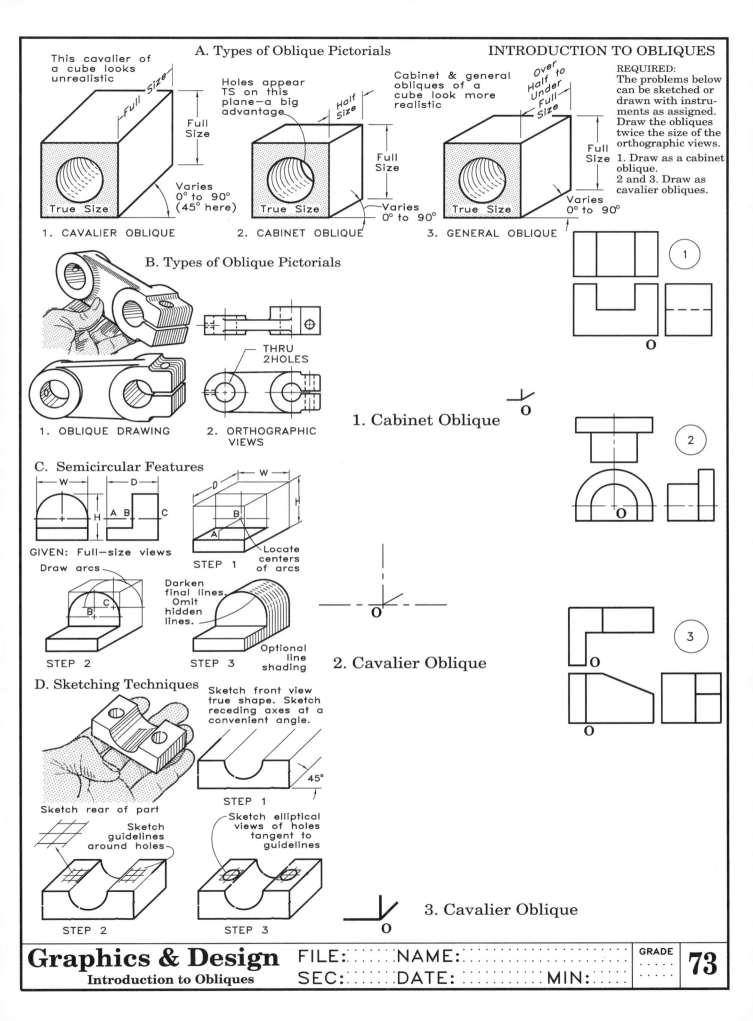

A. Types of Oblique Pictorials

This cavalier of a cube looks unrealistic

Full Size

Full Size

True Size

Varies 0° to 90° (45° here)

1. CAVALIER OBLIQUE

Holes appear TS on this plane—a big advantage

Half Size

Full Size

True Size

Varies 0° to 90°

2. CABINET OBLIQUE

Cabinet & general obliques of a cube look more realistic

Over Half to Under Full Size

Full Size

True Size

Varies 0° to 90°

3. GENERAL OBLIQUE

REQUIRED:
The problems below can be sketched or drawn with instruments as assigned. Draw the obliques twice the size of the orthographic views.

1. Draw as a cabinet oblique.
2 and 3. Draw as cavalier obliques.

B. Types of Oblique Pictorials

1. OBLIQUE DRAWING

THRU 2HOLES

2. ORTHOGRAPHIC VIEWS

C. Semicircular Features

W D

H A B C

GIVEN: Full—size views

D W H

B A

Locate centers of arcs

STEP 1

Draw arcs

C B

STEP 2

Darken final lines. Omit hidden lines.

Optional line shading

STEP 3

D. Sketching Techniques

Sketch front view true shape. Sketch receding axes at a convenient angle.

45°

STEP 1

Sketch rear of part

Sketch guidelines around holes

STEP 2

Sketch elliptical views of holes tangent to guidelines

STEP 3

○ 1

O

1. Cabinet Oblique

O

○ 2

O

2. Cavalier Oblique

O

○ 3

O

O

3. Cavalier Oblique

O

Graphics & Design
Introduction to Obliques

FILE: .... NAME: ....

SEC: .... DATE: .... MIN: ......

GRADE

73

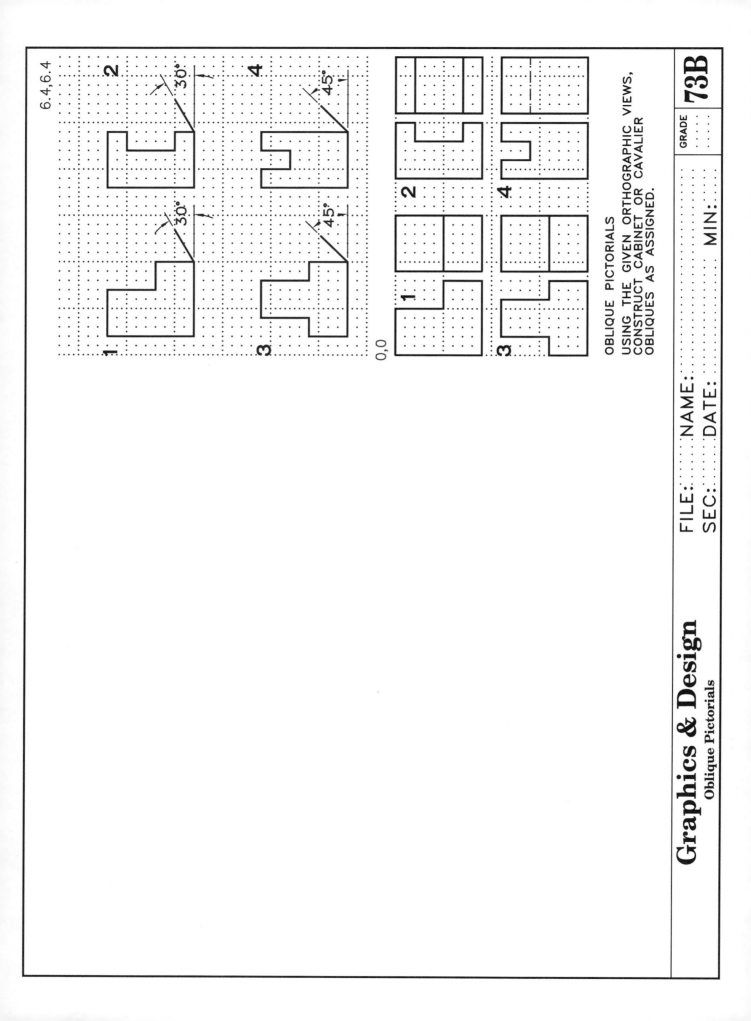

6.4, 6.4

1   2   30°
3   4   45°

0,0

OBLIQUE PICTORIALS

USING THE GIVEN ORTHOGRAPHIC VIEWS,
CONSTRUCT CABINET OR CAVALIER
OBLIQUES AS ASSIGNED.

1   2
3   4

Graphics & Design
Oblique Pictorials

FILE: .... NAME: ....
SEC: .... DATE: ....

GRADE **73B**
.... MIN: ....

PICTORIALS can be constructed by following the instructional examples given here. See page 50B.*

A. OBLIQUES: Sketch oblique views of four of the assigned problems on a single A-size (8.5 x 11) sheet.

B. ISOMETRICS: Sketch isometric views of four of the assigned problems on a single A-size (8.5 x 11) sheet.

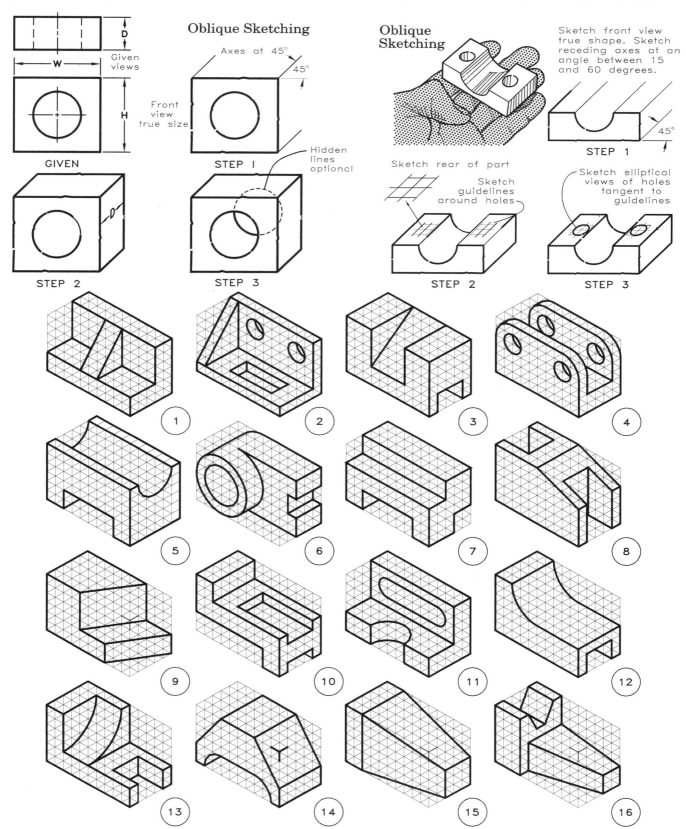

Oblique Sketching

Given views

D

W

H

GIVEN

Axes at 45°

45°

Front view true size

Hidden lines optional

STEP 1

STEP 2

D

STEP 3

Oblique Sketching

Sketch front view true shape. Sketch receding axes at an angle between 15 and 60 degrees.

45°

STEP 1

Sketch rear of part

Sketch guidelines around holes

STEP 2

Sketch elliptical views of holes tangent to guidelines

STEP 3

1 2 3 4

5 6 7 8

9 10 11 12

13 14 15 16

Graphics & Design
Oblique and Isometric Sketching

FILE: NAME:

SEC: DATE: MIN:

GRADE

74

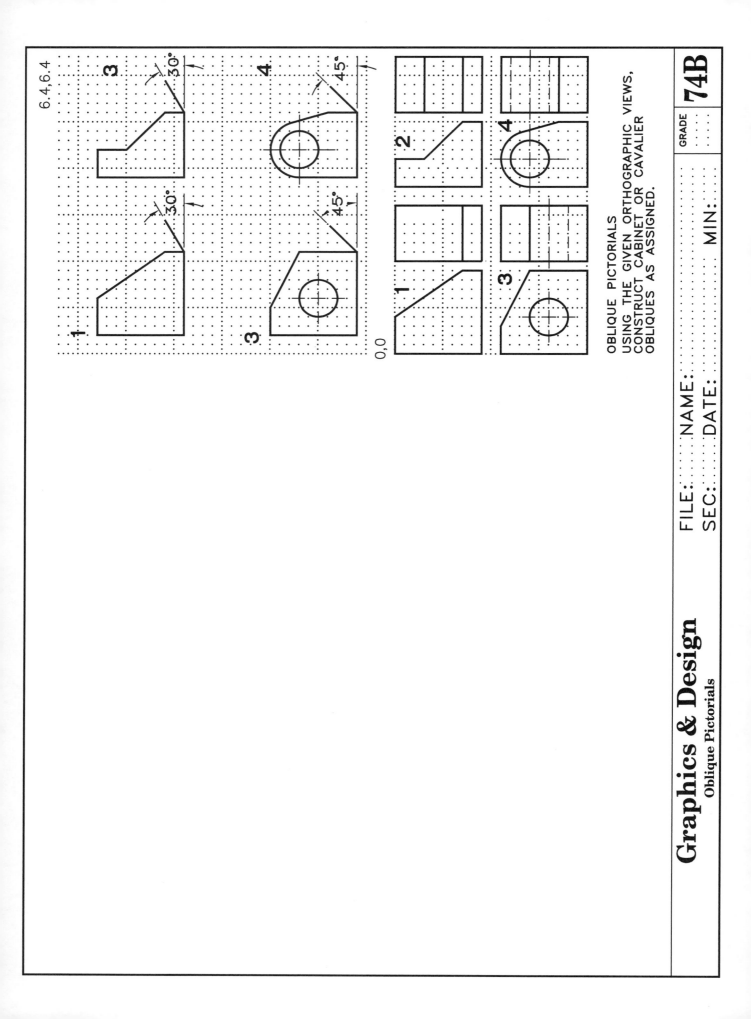

6.4, 6.4

OBLIQUE PICTORIALS
USING THE GIVEN ORTHOGRAPHIC VIEWS,
CONSTRUCT CABINET OR CAVALIER
OBLIQUES AS ASSIGNED.

Graphics & Design
Oblique Pictorials

FILE: NAME:
SEC: DATE: MIN:

GRADE 74B

Partial dimensions are given on these problems, but some are missing on all of them. You must determine all missing information as if you were the designer. Sketch your orthographic views on A-size sheets (8.5 x 11), either in a vertical or horizontal format. Consider how you would redesign each part that could function as well or better than the given configurations. See page 50B.

CORNER BRACKET

Make a three-view sketch of the corner bracket by using the given dimensions and estimating the others on a size A sheet. Label the views and give overall dimension H, W. and D. (ThomasNet)

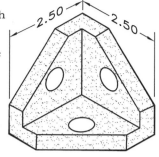

1.   CORNER BRACKET
    ALUMINUM
    16 REQUIRED

EYE BRACKET

Make a three-view sketch of the clevis bracket by using the given dimensions and estimating the others on a size A sheet. Label the views and give overall dimensions of H, W, and D. The hole that links with a clevis has a diameter of 26 mm. (ThomasNet)

2.   CLEVIS BRACKET
    ALUMINUM
    36 REQUIRED

ANGLE BRACKET

Make a three-view sketch of the bracket using the given dimensions and estimating others on a size A sheet. Label the views and give overall dimensions H, W, and D. Notice the section of the countersunk holes. (ThomasNet)

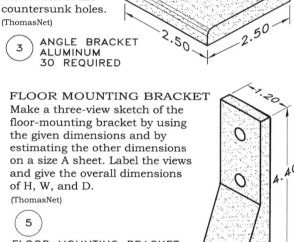

3.   ANGLE BRACKET
    ALUMINUM
    30 REQUIRED

EYE BRACKET

Make a three-view sketch of the eye bracket by using the given dimensions and estimating the others on a size A sheet. Label the views and give overall dimensions of H, W, and D. The hole through the boss has a diameter of .70 inches. (ThomasNet)

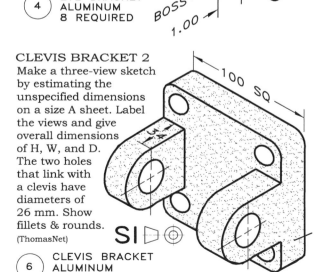

4.   EYE BRACKET
    ALUMINUM
    8 REQUIRED

FLOOR MOUNTING BRACKET

Make a three-view sketch of the floor-mounting bracket by using the given dimensions and by estimating the other dimensions on a size A sheet. Label the views and give the overall dimensions of H, W, and D. (ThomasNet)

5.

FLOOR MOUNTING BRACKET
ALUMINUM
24 REQUIRED

CLEVIS BRACKET 2

Make a three-view sketch by estimating the unspecified dimensions on a size A sheet. Label the views and give overall dimensions of H, W, and D. The two holes that link with a clevis have diameters of 26 mm. Show fillets & rounds. (ThomasNet)

6.   CLEVIS BRACKET
    ALUMINUM
    36 REQUIRED

PLAIN CLAMP:

Sketch the necessary views of the clamp to describe its details on a size A sheet using the approximate dimensions.

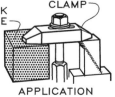

7.   PLAIN CLAMP
    STEEL—6 REQ
    APPROX SIZE:
    6"X 2"X 1" THICK

Partial dimensions are given on these problems, but some are missing on all of them. You must determine all missing information as if you were the designer. Sketch your orthographic views on A-size sheets (8.5 x 11), either in a vertical or horizontal format. Consider how you would redesign each part that could function as well or better than the given configurations. See 50B.

DESIGN 9: Sketch orthographic views of the parts of the hinge assembly on a size A sheet. Consider designing a better one. Estimate missing dimensions.

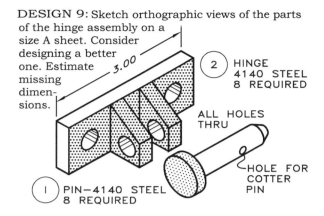

3.00

② HINGE
4140 STEEL
8 REQUIRED

ALL HOLES
THRU

① PIN—4140 STEEL
8 REQUIRED

HOLE FOR
COTTER
PIN

DESIGN 8: Sketch the orthographic views of the handle on a size A sheet. Add a threaded hole at the bent end for a set screw to secure the square shaft to it.

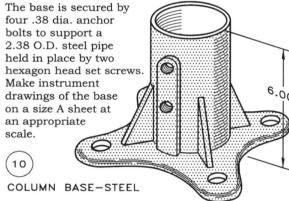

SHAFT

1.00

.32 SQUARE
HOLE

Ø.30

.75

2.75

⑧ HANDLE—STEEL
22 REQUIRED

DESIGN 10: COLUMN BASE

The base is secured by four .38 dia. anchor bolts to support a 2.38 O.D. steel pipe held in place by two hexagon head set screws. Make instrument drawings of the base on a size A sheet at an appropriate scale.

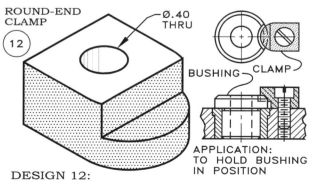

6.00

⑩

COLUMN BASE—STEEL

DESIGN 11: Move the flange to the center of the part and show 6 holes instead of 4 and add 2 ribs on each side. Sketch the necessary views to describe the the new design. Label your sketches as needed.

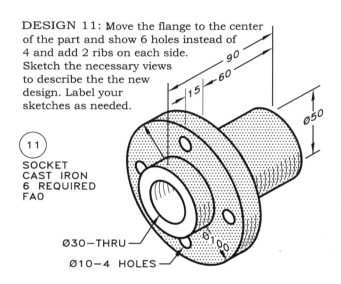

90

15

60

Ø50

⑪

SOCKET
CAST IRON
6 REQUIRED
FAO

Ø30—THRU

Ø10—4 HOLES

ROUND-END CLAMP

⑫

Ø.40
THRU

BUSHING

CLAMP

APPLICATION:
TO HOLD BUSHING
IN POSITION

DESIGN 12:

Sketch the necessary views of the clamp on a size A sheet to describe its details. Sketch it as a pictorial as well and label all views.

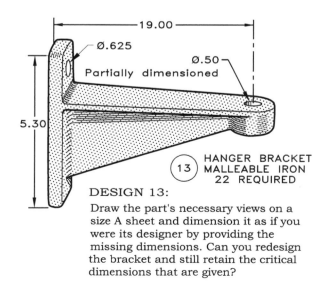

19.00

Ø.625

Partially dimensioned

Ø.50

5.30

HANGER BRACKET
MALLEABLE IRON
22 REQUIRED

⑬

DESIGN 13:

Draw the part's necessary views on a size A sheet and dimension it as if you were its designer by providing the missing dimensions. Can you redesign the bracket and still retain the critical dimensions that are given?

Graphics & Design
Orthographic Design Sketching

FILE:

SEC:

NAME:

DATE:

MIN:

GRADE

75B

A. LINE TYPES AND WEIGHTS $\frac{1}{16}$ — $\frac{1}{8}$ — 1

By following the examples at the left, draw double-size views of the given parts using your dividers to transfer the dimensions. Apply the differences in line weights.

Visible Line

Centerline

$\frac{1}{32}$ — $\frac{1}{8}$

Hidden Line

Break Line

$\frac{1}{32}$ — $\frac{1}{8}$ — 1

Cutting Plane

Long-Break Line

B. THE COMPASS
is for drawing circles that are too big for the circle template. Begin by drawing the centerlines first and make several circles with slightly different radii to make the circle's lineweight the appropriate thickness.

1.

LINK

SHARPEN ON OUTSIDE

2.

Sandpaper

Slight angle

Slight angle

POINT LENGTHS

$\frac{3}{8}$

BEVEL POINT

COVER PIECE

C. THE CIRCLE TEMPLATE
is used for drawing small circles by aligning the cross marks with the circle's centerlines on the drawing.

$\frac{3}{4}$

Diameter of circle on the template

3.

PIVOT BEARING

Break line

D. THE DIVIDERS
are helpful in transferring measurements from a scale to a location on a drawing.

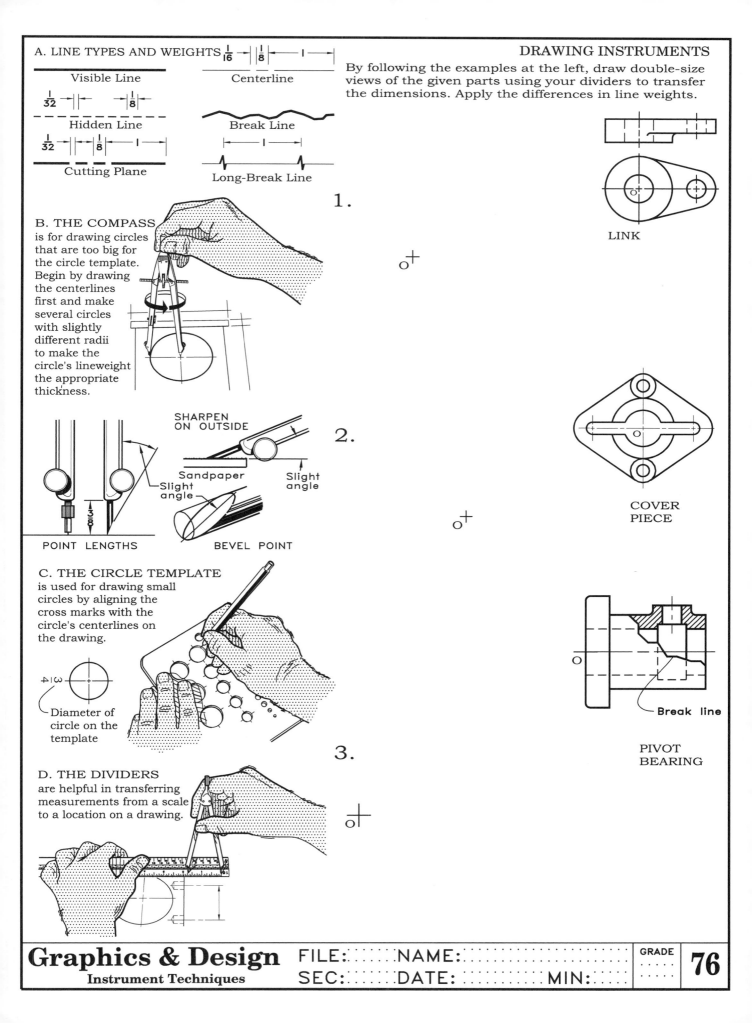

Graphics & Design
Instrument Techniques

FILE: .... NAME: ....

SEC: .... DATE: .... MIN: ....

GRADE

76

6.4,6.4

0,0

ORTHOGRAPHIC PROJECTION

PLOT THE VIEWS OF THE PART AND
ADD THE MISSING LINES. LINES MAY BE
MISSING IN ALL VIEWS.

Graphics & Design
Oblique Pictorials

FILE: NAME:
SEC: DATE: MIN:

GRADE **76B**

A. THE DIVIDERS
are used to divide a line into equal segments or to transfer dimensions from one part of a drawing to another location.

B. THE DIVIDERS
are helpful in transferring measurements from a scale to a location on a drawing.

C. THE BOW DIVIDERS
are used for transferring measurements with greater accuracy because the settings are held by the adjusting screw.

Adjusting wheel

Threaded screw

D. THE COMPASS
is for drawing circles that are too big for the circle template. Begin by drawing the centerlines first and make several circles with slightly different radii to make the circle's lineweight the appropriate thickness.

E. LEAD SHARPENING

SHARPEN ON OUTSIDE

Sandpaper

Slight angle

Slight angle

POINT LENGTHS

BEVEL POINT

PROBLEMS 1-3: Use your drawing instruments as described in the examples at the left, complete the three-view orthographic drawings of the parts. Show the centerlines for the circular holes.

1. LINK

2. CONNECTOR

3. CLEVIS

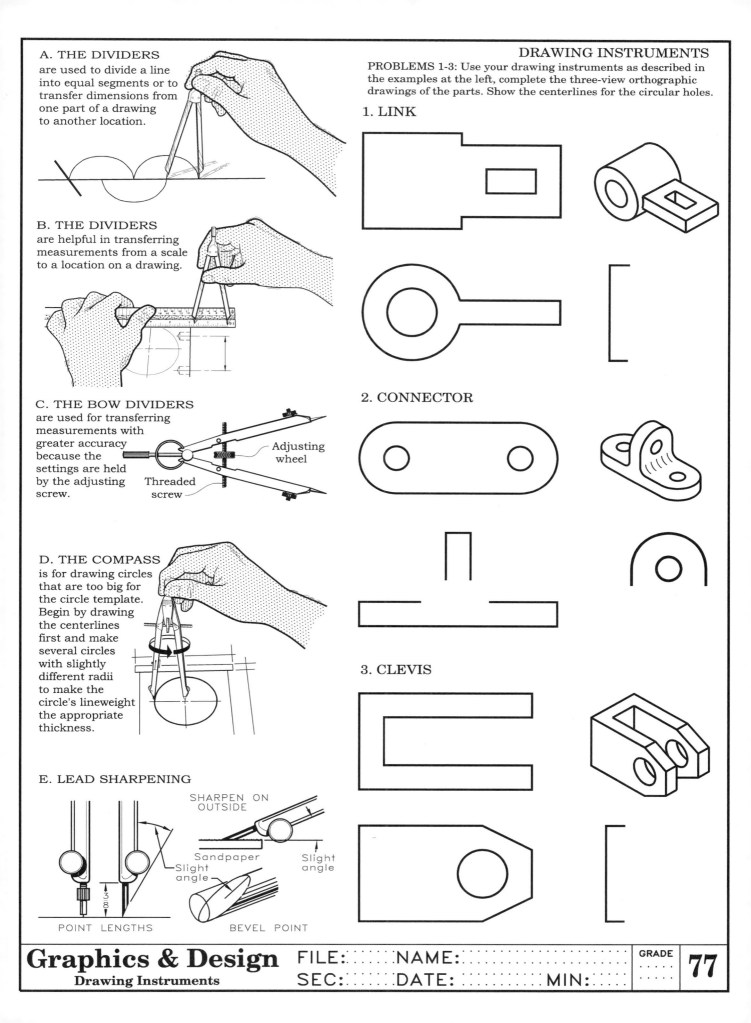

Graphics & Design
Drawing Instruments

FILE: . . . . NAME: . . . .
SEC: . . . . DATE: . . . . MIN: . . . . .

GRADE

77

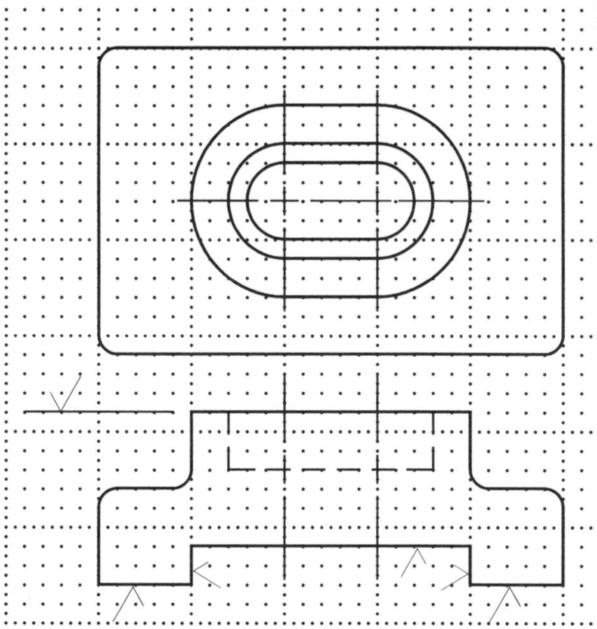

6.4,6.4

0,0

ORTHOGRAPHIC PROJECTION

PLOT THE TWO VIEWS OF THE PART.
ADD THE MISSING LINES WHICH MAY BE
MISSING IN BOTH VIEWS.

Graphics & Design
Orthographic Projection

FILE: NAME:
SEC: DATE:

GRADE 77B
MIN:

Complete the missing views of each of the parts with
instruments. Give dimensions W, D and H and labels.

1. THIS SINGLE SLOPING PLANE
appears as an edge in the front view and
foreshortened in the top and side views.

DIVIDERS must be used to
transfer the dimensions
of the parts to the space at
the right.

TOP VIEW   SAMPLE

D

W   D

H

FRONT VIEW   R SIDE V

4. LIFTING PIECE

2. THREE SLOPING PLANES
appear as edges in top and front views, but as
foreshortened in the side view.

5. CENTERING BLOCK

3. A COMPOUND SLOPING PLANE
does not appear as an edge in any of the views, but
as foreshortened in all three views.

6. CORNER GUIDE

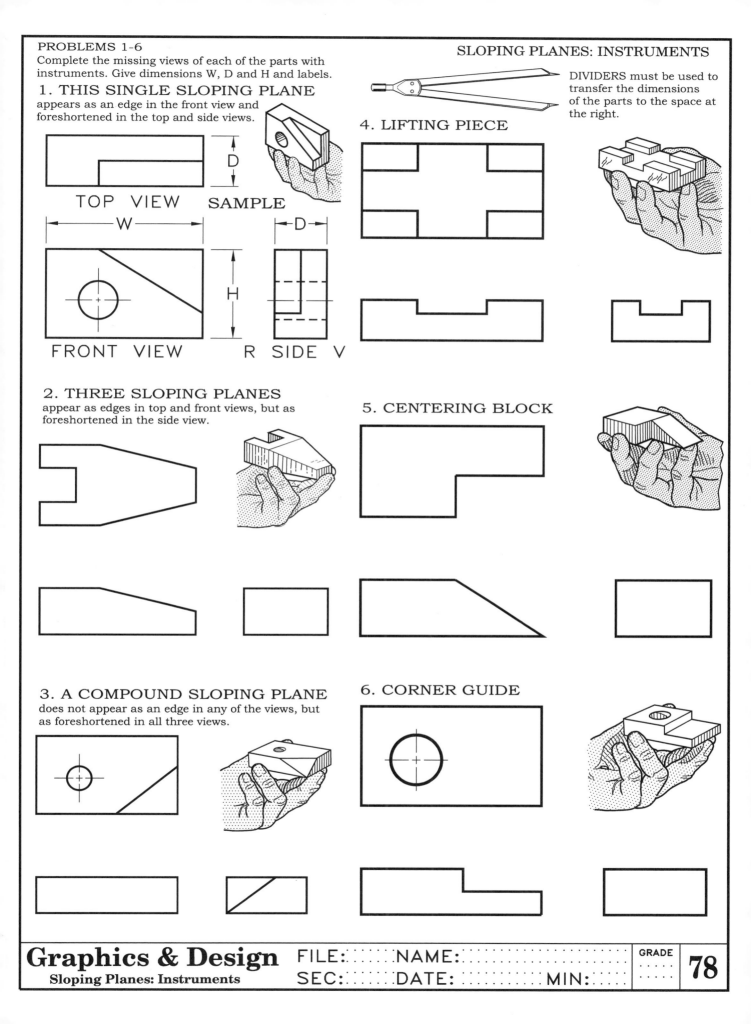

Graphics & Design
Sloping Planes: Instruments

FILE: · · · · · · NAME: ·
SEC: · · · · · · DATE: · · · · · · · · · · MIN: · · · · · · · ·

GRADE · · · ·

78

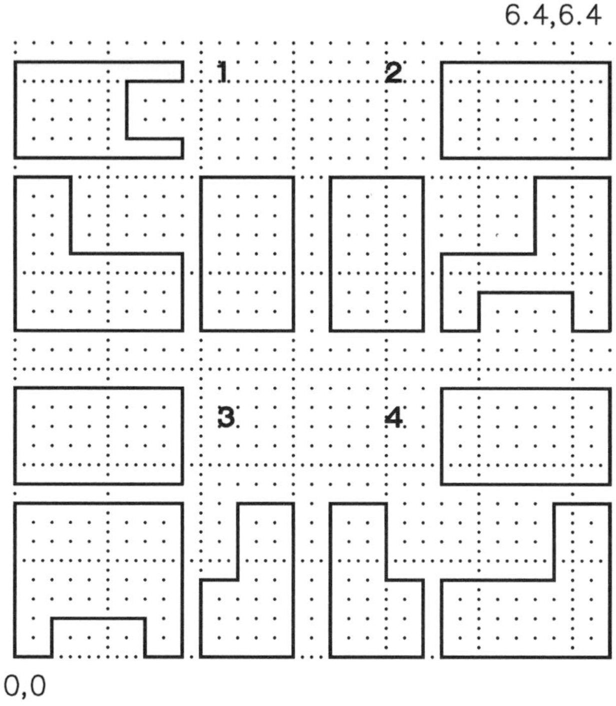

6.4,6.4

1   2   3   4

0,0

ORTHOGRAPHIC PROJECTION

PLOT THE VIEWS OF THE FOUR PARTS.
ADD THE MISSING LINES WHICH MAY BE
MISSING IN ALL VIEWS.

Graphics & Design
Orthographic Projection

FILE: .... NAME: ....
SEC: .... DATE: ....

GRADE ....

78B

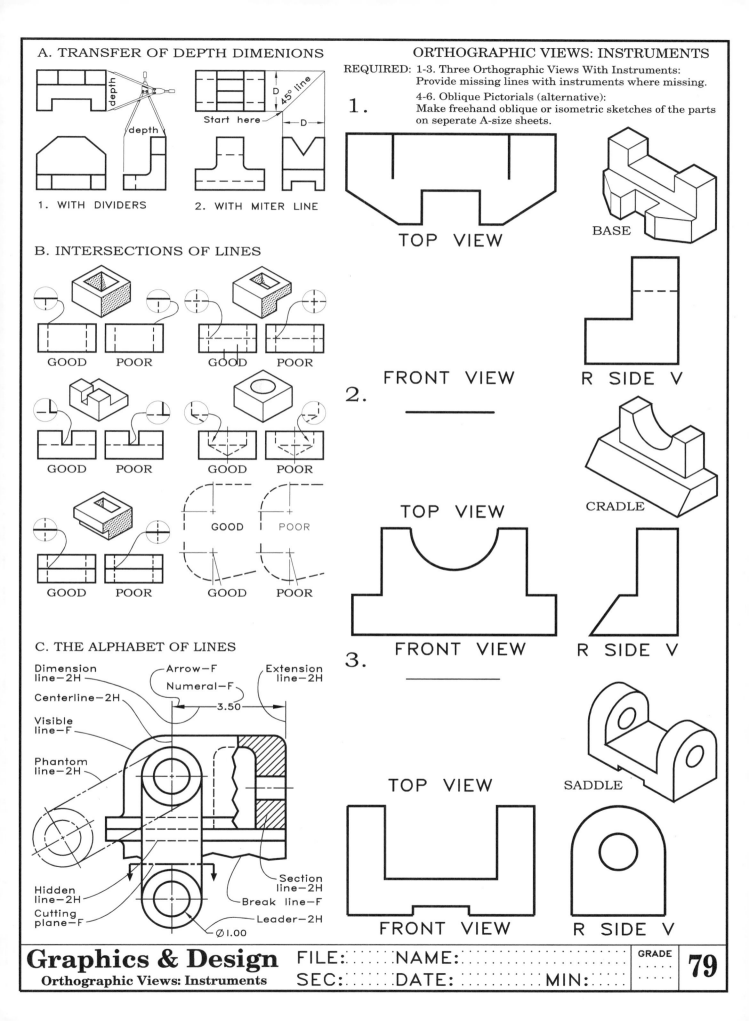

A. TRANSFER OF DEPTH DIMENIONS

depth

depth

1. WITH DIVIDERS

D

45° line

Start here

D

2. WITH MITER LINE

B. INTERSECTIONS OF LINES

GOOD   POOR   GOOD   POOR

GOOD   POOR   GOOD   POOR

GOOD   POOR   GOOD   POOR

C. THE ALPHABET OF LINES

Dimension line—2H
Centerline—2H
Visible line—F
Phantom line—2H
Arrow—F
Numeral—F
3.50
Extension line—2H
Hidden line—2H
Cutting plane—F
Section line—2H
Break line—F
Leader—2H
Ø1.00

ORTHOGRAPHIC VIEWS: INSTRUMENTS

REQUIRED: 1-3. Three Orthographic Views With Instruments:
Provide missing lines with instruments where missing.

4-6. Oblique Pictorials (alternative):
Make freehand oblique or isometric sketches of the parts on seperate A-size sheets.

1.

TOP VIEW

BASE

FRONT VIEW _____   R SIDE V

2.

TOP VIEW

CRADLE

FRONT VIEW   R SIDE V

3. _____

TOP VIEW

SADDLE

FRONT VIEW   R SIDE V

6.4,6.4

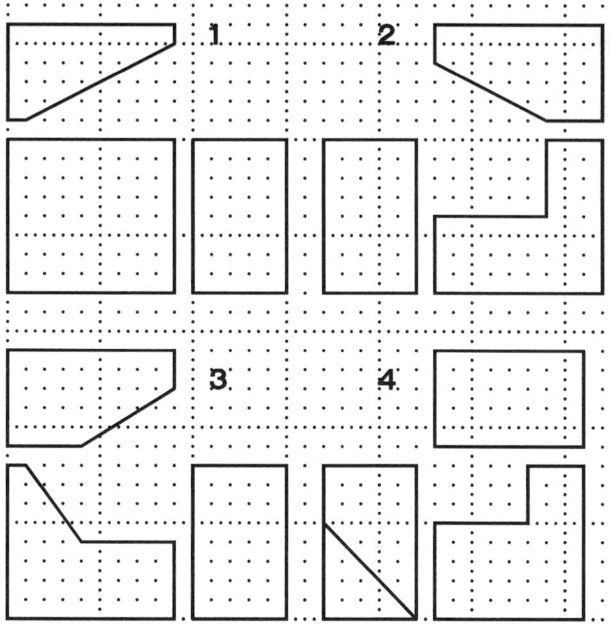

2

1

4

3

0,0

ORTHOGRAPHIC PROJECTION

PLOT THE VIEWS OF THE FOUR PARTS.
ADD THE MISSING LINES WHICH MAY BE
MISSING IN ALL VIEWS.

Graphics & Design
Orthographic Projection

FILE:
NAME:
SEC:
DATE:
MIN:

A. THE SIX PRINCIPAL VIEWS

Six is the maximum number of principal planes.

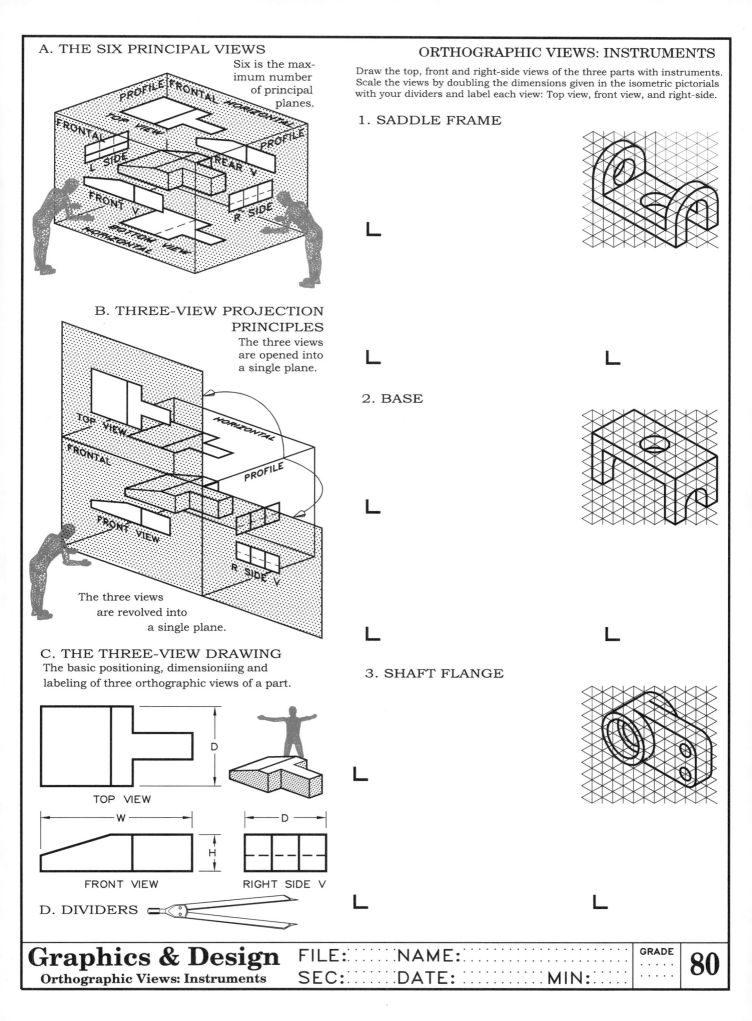

B. THREE-VIEW PROJECTION PRINCIPLES

The three views are opened into a single plane.

The three views are revolved into a single plane.

C. THE THREE-VIEW DRAWING

The basic positioning, dimensioniing and labeling of three orthographic views of a part.

TOP VIEW

FRONT VIEW

RIGHT SIDE V

D. DIVIDERS

ORTHOGRAPHIC VIEWS: INSTRUMENTS

Draw the top, front and right-side views of the three parts with instruments. Scale the views by doubling the dimensions given in the isometric pictorials with your dividers and label each view: Top view, front view, and right-side.

1. SADDLE FRAME

2. BASE

3. SHAFT FLANGE

Graphics & Design
Orthographic Views: Instruments

FILE: ......NAME: ....

SEC: ....DATE: ....MIN: ......

GRADE

80

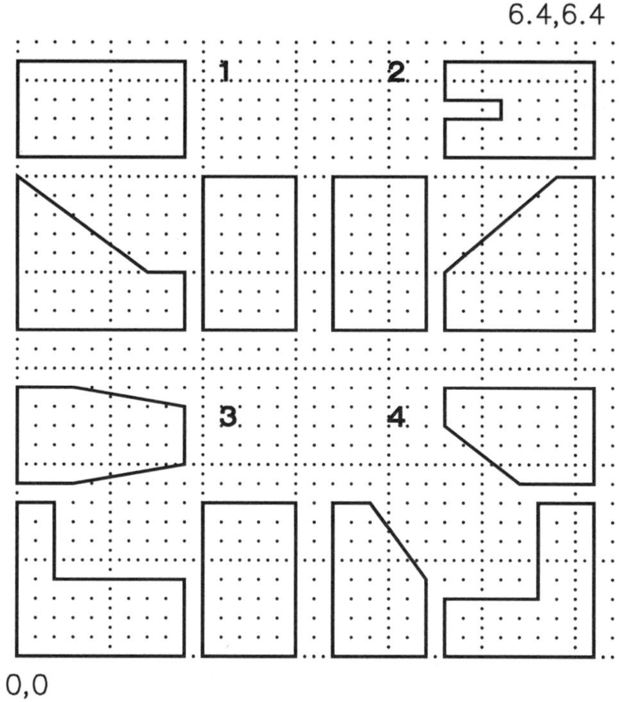

6.4,6.4

1   2

3   4

0,0

ORTHOGRAPHIC PROJECTION

PLOT THE VIEWS OF THE FOUR PARTS.
ADD THE MISSING LINES WHICH MAY BE
MISSING IN ALL VIEWS.

Graphics & Design
Orthographic Projection

FILE: .... NAME: ....
SEC: .... DATE: ....

GRADE 80B

MIN: ....

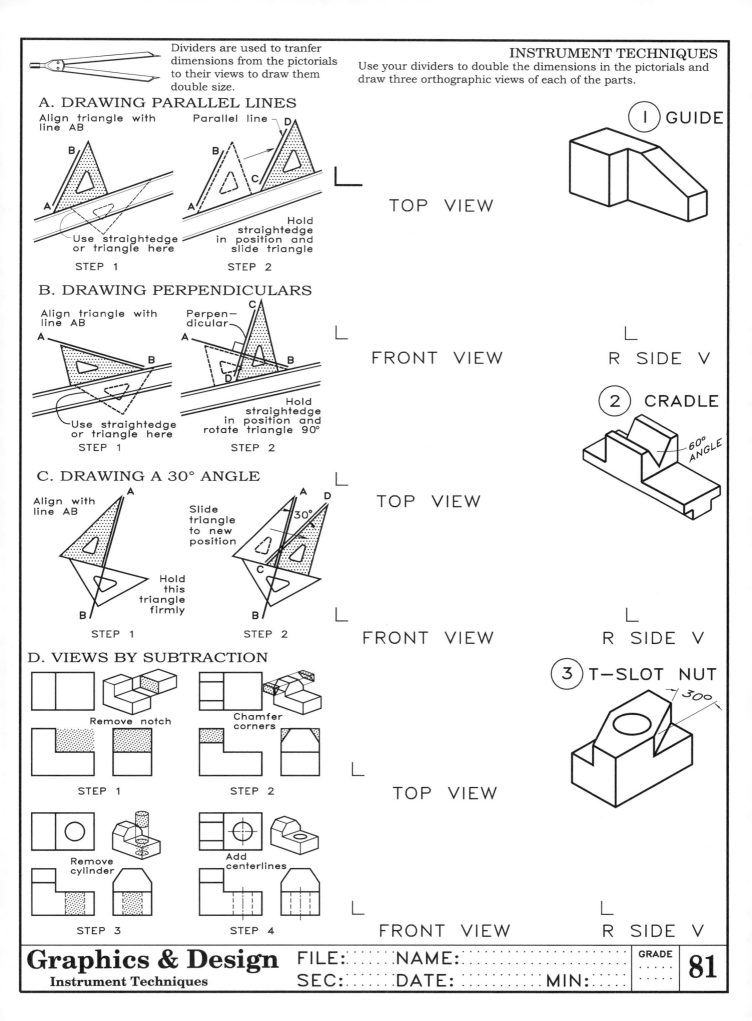

Dividers are used to tranfer dimensions from the pictorials to their views to draw them double size.

Use your dividers to double the dimensions in the pictorials and draw three orthographic views of each of the parts.

A. DRAWING PARALLEL LINES

Align triangle with line AB

Parallel line

B

D

A

V

C

Use straightedge or triangle here

Hold straightedge in position and slide triangle

STEP 1   STEP 2

B. DRAWING PERPENDICULARS

Align triangle with line AB

A

B

Perpen- dicular

C

B

D

Use straightedge or triangle here

Hold straightedge in position and rotate triangle 90°

STEP 1   STEP 2

C. DRAWING A 30° ANGLE

Align with line AB

A

B

Slide triangle to new position

A

D

30°

C

B

Hold this triangle firmly

STEP 1   STEP 2

D. VIEWS BY SUBTRACTION

Remove notch

Chamfer corners

STEP 1   STEP 2

Remove cylinder

Add centerlines

STEP 3   STEP 4

① GUIDE

TOP VIEW

FRONT VIEW   R SIDE V

② CRADLE

60° ANGLE

TOP VIEW

FRONT VIEW   R SIDE V

③ T—SLOT NUT

300

TOP VIEW

FRONT VIEW   R SIDE V

Graphics & Design
Instrument Techniques

FILE: NAME:

SEC: DATE: MIN:

GRADE

81

6.4,6.4

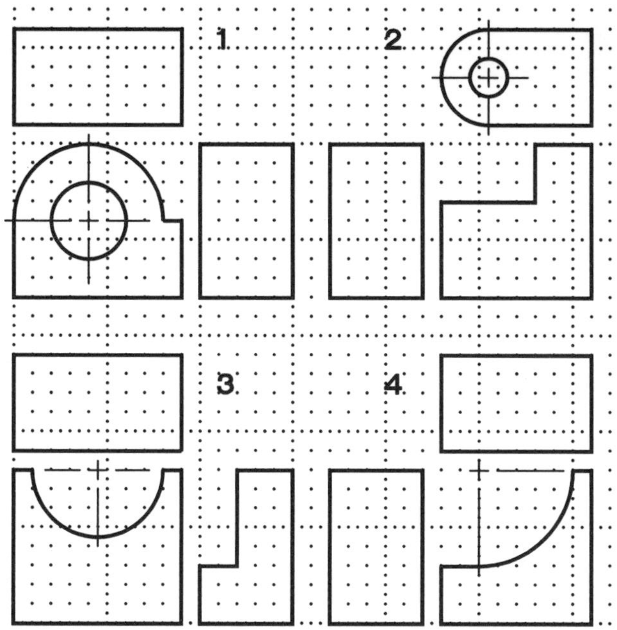

2

1

4

3

0,0

ORTHOGRAPHIC PROJECTION

PLOT THE VIEWS OF THE FOUR PARTS.
ADD THE MISSING LINES WHICH MAY BE
MISSING IN ALL VIEWS.

Graphics & Design
Orthographic Projection

FILE: .... NAME: ....

SEC: .... DATE: ....

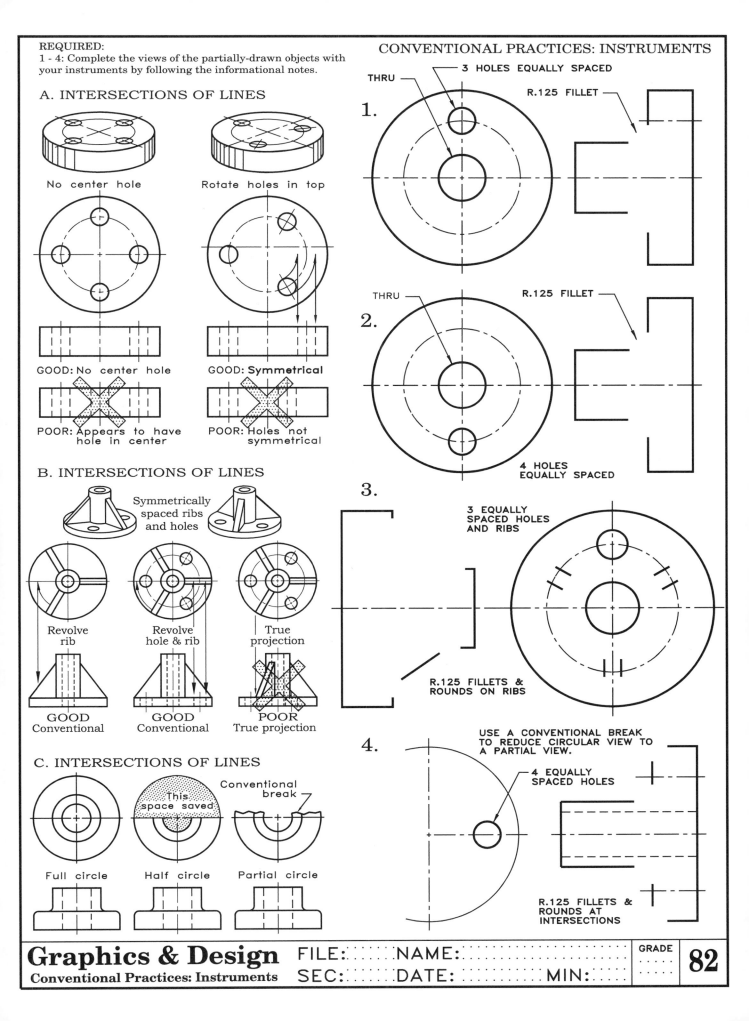

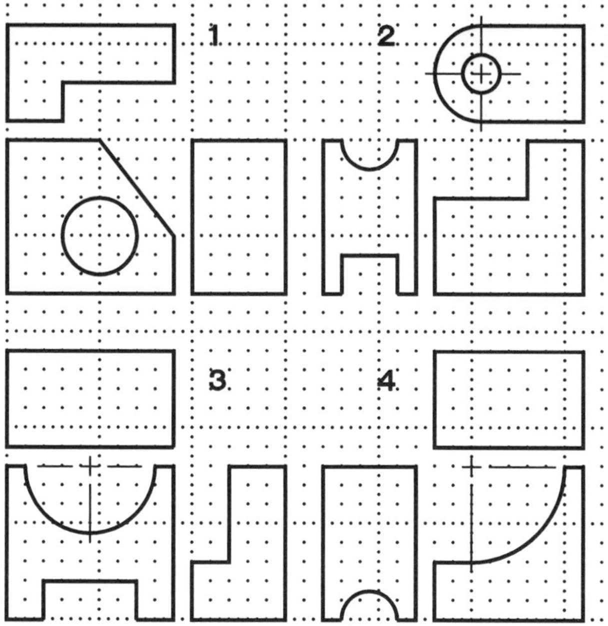

6.4,6.4

2

1

4

3

0,0

ORTHOGRAPHIC PROJECTION

PLOT THE VIEWS OF THE FOUR PARTS.
ADD THE MISSING LINES WHICH MAY BE
MISSING IN ALL VIEWS.

Graphics & Design
Orthographic Projection

FILE: .... NAME: ....
SEC: .... DATE: ....

GRADE 82B
MIN: ....

A. ELLIPSE PROJECTION

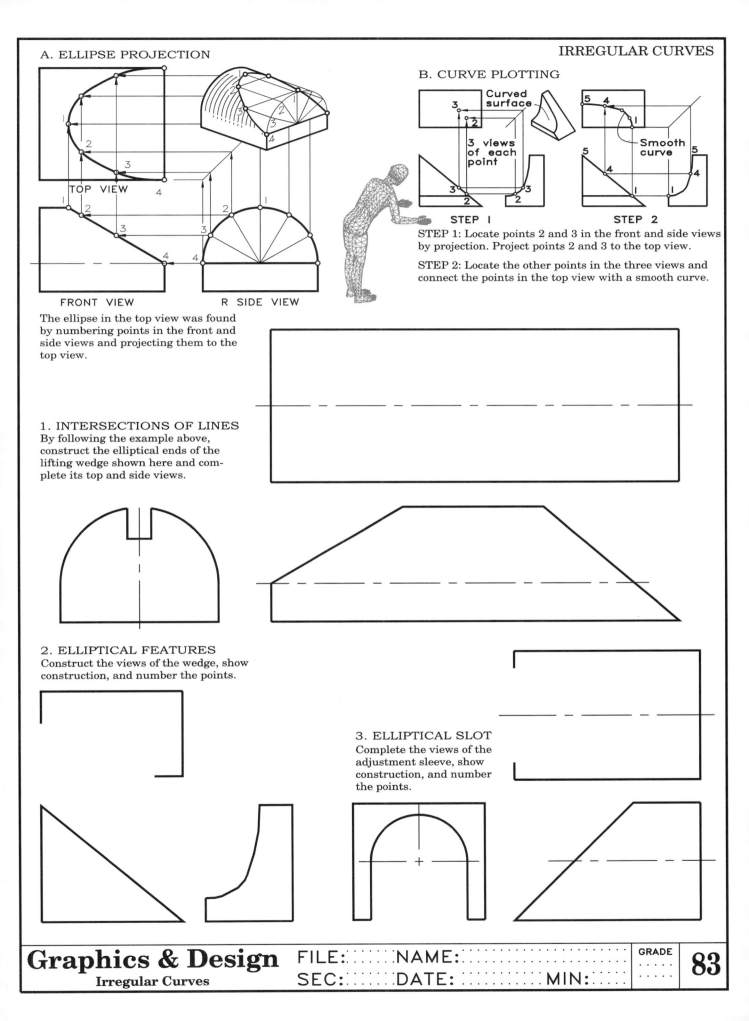

TOP VIEW

FRONT VIEW

R SIDE VIEW

The ellipse in the top view was found by numbering points in the front and side views and projecting them to the top view.

B. CURVE PLOTTING

Curved surface

3 views of each point

STEP 1

Smooth curve

STEP 2

STEP 1: Locate points 2 and 3 in the front and side views by projection. Project points 2 and 3 to the top view.

STEP 2: Locate the other points in the three views and connect the points in the top view with a smooth curve.

1. INTERSECTIONS OF LINES
By following the example above, construct the elliptical ends of the lifting wedge shown here and complete its top and side views.

2. ELLIPTICAL FEATURES
Construct the views of the wedge, show construction, and number the points.

3. ELLIPTICAL SLOT
Complete the views of the adjustment sleeve, show construction, and number the points.

Graphics & Design
Irregular Curves

FILE: NAME:
SEC: DATE: MIN:

GRADE

83

6.4,6.4

0,0

ORTHOGRAPHIC PROJECTION

PLOT THE THREE VIEWS OF THE PART.
ADD THE MISSING LINES WHICH MAY BE
MISSING IN ALL VIEWS.

Graphics & Design
Orthographic Views

FILE:........NAME:..................
SEC:.........DATE:..................

GRADE 83B
MIN:.....

A. FINISHED SURFACES

60°

Finish mark — Edge — Letter height

Rounds

Sharp corners

TOP VIEW

FRONT VIEW

Removed surface

FRONT VIEW

FRONT VIEW

B. FILLETS & ROUNDS

Fillet

Round

Finished surfaces

Fillet

Round

Sharp corners

C. FINISH MARKS: Apply to edge views

H=Letter height
Draw with F pencil

Place finish marks on all finished surfaces, visible or hidden, in the edge views

Specifications

1.6
0.8

60°

3H

1.5H

H

H

1.5H

A.   B.   C.   D.

D. FINISH ALL OVER -- FAO

FAO means "Finish All Over."

This view is unnecessary

15

6 DIA

11

FRONT VIEW   FAO   R SIDE V

---

1. BRACKET
Draw the front view of the bracket to show visible, hidden, center-lines, and finish marks.
METRIC SCALE: 1:20

R22   Ø20-2 HOLES   SI

16

72

50

38

R6

R12   20   26   12

76

FRONT VIEW

16   12

1   BRACKET
    CAST IRON

42   76   O

2. FINISH MARKS
Duplicate the finish marks given in the example located at points A thru D, assuming the letter height (H) is 3.5 mm.

A.   B.   C.   D.

3. FINISH MARKS: ALL OVER
Draw the part triple size as an orthographic view with dimensions.

Graphics & Design
Finish Marks & Single Views

FILE: .... NAME: ....

SEC: .... DATE: .... MIN: ......

GRADE

84

6.4,6.4

0,0

ORTHOGRAPHIC PROJECTION

DRAW THE TWO VIEWS OF THE PART
AND ADD ANY LINES THAT MAY BE
MISSING.

Graphics & Design
Orthographic Projection

FILE: .... NAME: ....

SEC: .... DATE: ....

GRADE 84B

MIN: ....

DESIGN 1: BASE FLANGE

Make an instrument drawing of the flange by using the general specifications and your judgment as its designer. Draw details in the space below.

2 SOCKET HEAD SCREWS

Ø1.90 THRU

Chamfer upper surface

Ø.375 THRU COUNTERSINK 4 HOLES

No fillet here

3.60 SQ base

Fillet—All around

Round upper edge—4 sides

Height=2.40

① BASE FLANGE 1030 STEEL—1 REQ

ORTHOGRAPHIC VIEWS: DESIGN
DESIGN 2: FLANGE BEARING

Modify as follows: Add 4 bosses (.20 thick) at each corner and round each corner concentric with each hole. Draw orthographic views with instruments of the modified part in the space below

Ø2.40 THRU

Ø3.20

Ø.80 4 HOLES

.60

2.40

BOSS

6.00SQ

4.00SQ

② FLANGE BEARING CAST IRON 4 REQUIRED

REQUIRED:
Complete the problem assigned (Design 1 or Design 2) and follow the instruction given for each of them. Refer to sheet 50B.*

GENERAL INSTRUCTIONS:

Draw the orthographic views of the one assigned from those below using instruments on an A-size sheet. Complete the title block for each with the information assigned by your instructor, or use the format given on this sheet. Strive for precision and line quality. Refer to sheet 50B.*

CASTLE INSERT

Draw two full-size orthographic views of the insert on a size A sheet and show the overall dimensions of H, W, and D and label the views. Specify rounds at the top edges of the base and at the base of the cone.

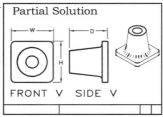

Partial Solution

FRONT V   SIDE V

1.00–8UNC–2B
2.00 DEEP

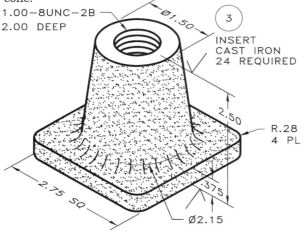

Ø1.50

3

INSERT
CAST IRON
24 REQUIRED

2.50

R.28
4 PL

2.75 SQ

.375

Ø2.15

ROD GUIDE

Draw full-size orthographic views of the top, front, and side views of the rod guide. Give the overall dimensions of W, D, and H; show finish marks where appropriate.

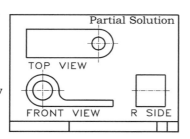

Partial Solution

TOP VIEW

FRONT VIEW   R SIDE

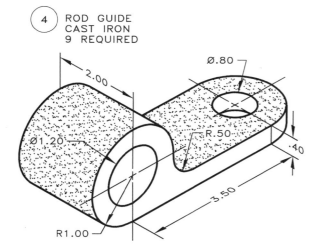

4   ROD GUIDE
CAST IRON
9 REQUIRED

2.00

Ø.80

R.50

Ø1.20

.40

3.50

R1.00

CABLE CLAMP

Draw triple-size orthographic views of the top, front and side views of the clamp on a size A sheet. Give H, W, and D dimensions and label the views.

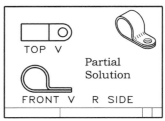

TOP V

Partial Solution

FRONT V   R SIDE

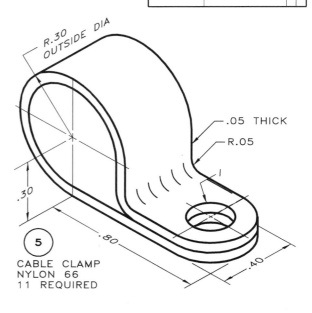

R.30
OUTSIDE DIA

.05 THICK

R.05

.30

.80

.40

5

CABLE CLAMP
NYLON 66
11 REQUIRED

PACKAGE BRACKET

Draw full-size orthographic views of two views of the package bracket on a size A sheet. Give overall dimensions of W, D, and H. Give appropriate fillets and rounds at the bends.

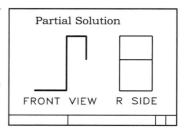

Partial Solution

FRONT VIEW   R SIDE

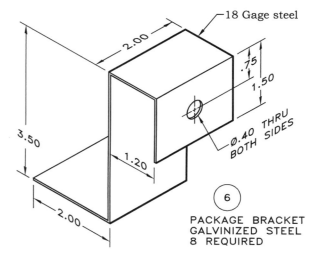

18 Gage steel

2.00

.75

1.50

3.50

Ø.40 THRU
BOTH SIDES

1.20

2.00

6

PACKAGE BRACKET
GALVINIZED STEEL
8 REQUIRED

Graphics & Design
Orthographic Applications

FILE: . . . . NAME: . . . .

SEC: . . . . . DATE: . . . . MIN: . . . . .

GRADE

85B

The example below shows the steps in finding the true-size view of a plane that appears as an edge in the top view.

Step 1: Establish viewpoint

Draw H1 parallel to edge of plane.

Ref. Line parallel to edge

Auxiliary Pl.

Sight

Edge

TOP

FRONT

Transfer H with dividers.

Step 2: Transfer measurements

Transfer H dimension from HF to H1 plane.

TS

Sight H

Parallel

TOP

FRONT

Transfer H with dividers.

Step 3: Connect points

Transfer other H dimensions to complete the auxiliary view.

TOP

TS

AUX. V

FRONT

Transfer H with dividers.

1.

SIGHT

TOP VIEW

$\dfrac{H}{F}$

FRONT V

2.

At the base

TOP VIEW

HRP

HRP

FRONT V

PROBLEMS 1 - 4:
Using instruments, draw the auxiliary views indicated by the lines of sight to show the inclined surfaces true size.

3.

TOP VIEW

HRP

FRONT V

HRP

Horiz. Reference Plane

HRP

Measure on both sides of the HRP.

Through the part

*HRP (Horizontal Reference Plane)

4.

TOP VIEW

HRP

FRONT V

Graphics & Design
Auxiliary Views From the Top View

FILE: .... NAME: ....

SEC: .... DATE: .... MIN: ....

GRADE **86**

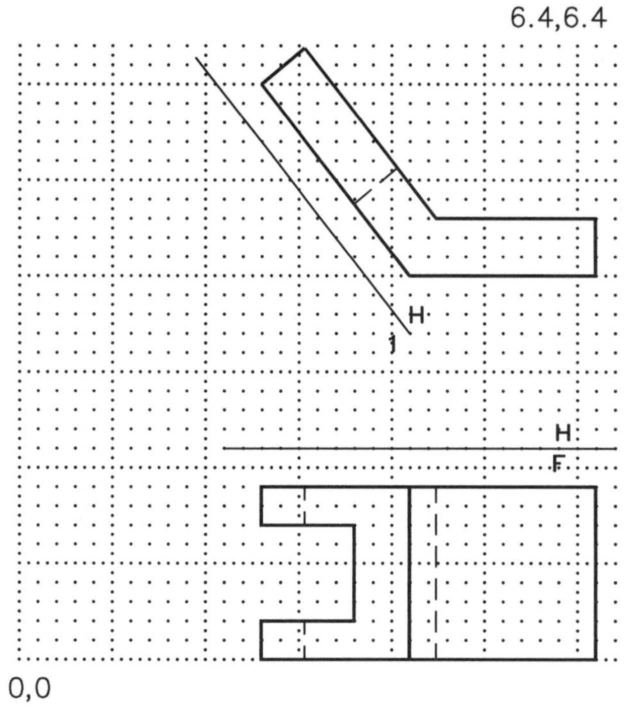

6.4,6.4

H
1

H
F

0,0

PRIMARY AUXILIARY VIEWS

DRAW THE GIVEN VIEWS AND PLOT AN
AUXILIARY VIEW OF THE INCLINED
SURFACE ONLY. LABEL THE REFERENCE
LINES AND NUMBER THE CORNERS OF
THE AUXILIARY VIEW.

Graphics & Design
Primary Auxiliary Views

FILE: .... NAME: ....
SEC: .... DATE: ....

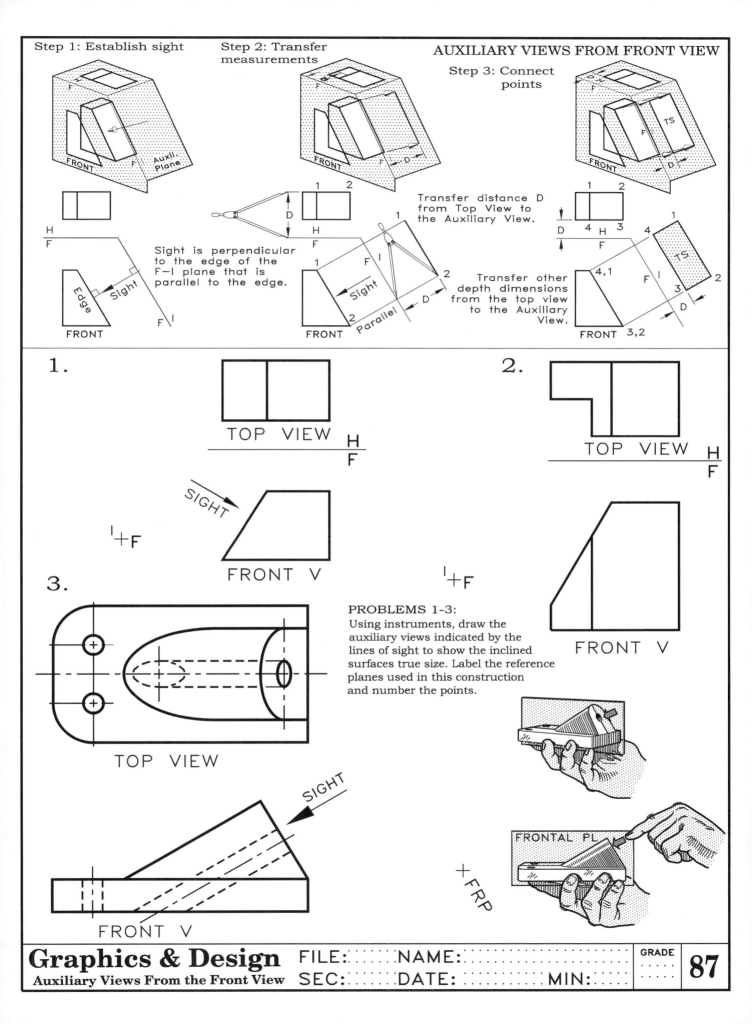

AUXILIARY VIEWS FROM FRONT VIEW

Step 1: Establish sight

Step 2: Transfer measurements

Step 3: Connect points

Sight is perpendicular to the edge of the F—I plane that is parallel to the edge.

Transfer distance D from Top View to the Auxiliary View.

Transfer other depth dimensions from the top view to the Auxiliary View.

1.

TOP VIEW

SIGHT

FRONT V

2.

TOP VIEW

FRONT V

3.

TOP VIEW

SIGHT

FRONT V

PROBLEMS 1-3:
Using instruments, draw the auxiliary views indicated by the lines of sight to show the inclined surfaces true size. Label the reference planes used in this construction and number the points.

FRONTAL PL

+ FRP

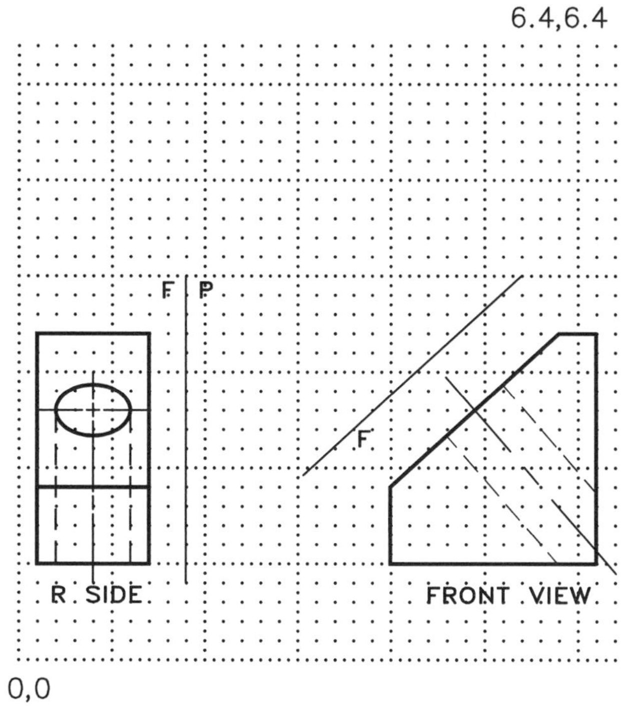

6.4,6.4

FRONT VIEW

F.

F | P
F | P

R SIDE

0,0

PRIMARY AUXILIARY VIEWS

DRAW THE GIVEN VIEWS AND PLOT AN
AUXILIARY VIEW OF THE INCLINED
SURFACE ONLY. LABEL THE REFERENCE
LINES AND NUMBER THE CORNERS OF
THE AUXILIARY VIEW.

Graphics & Design
Primary Auxiliary Views

FILE: · · · · · · NAME: · · · · · · · · · · · ·
SEC: · · · · · · DATE: · · · · · · MIN: · · · · · ·

Step 1: Establish sight

Sight is perpendicular to the edge view of the inclined plane. The auxiliary plane (P1) is parallel to the edge view of the inclined surface. Number the points as you progress.

Step 2: Transfer width measurements

Transfer W from Front View to the auxiliary view.

Step 3: Connect points

Transfer other W dimensions to auxiliary view.

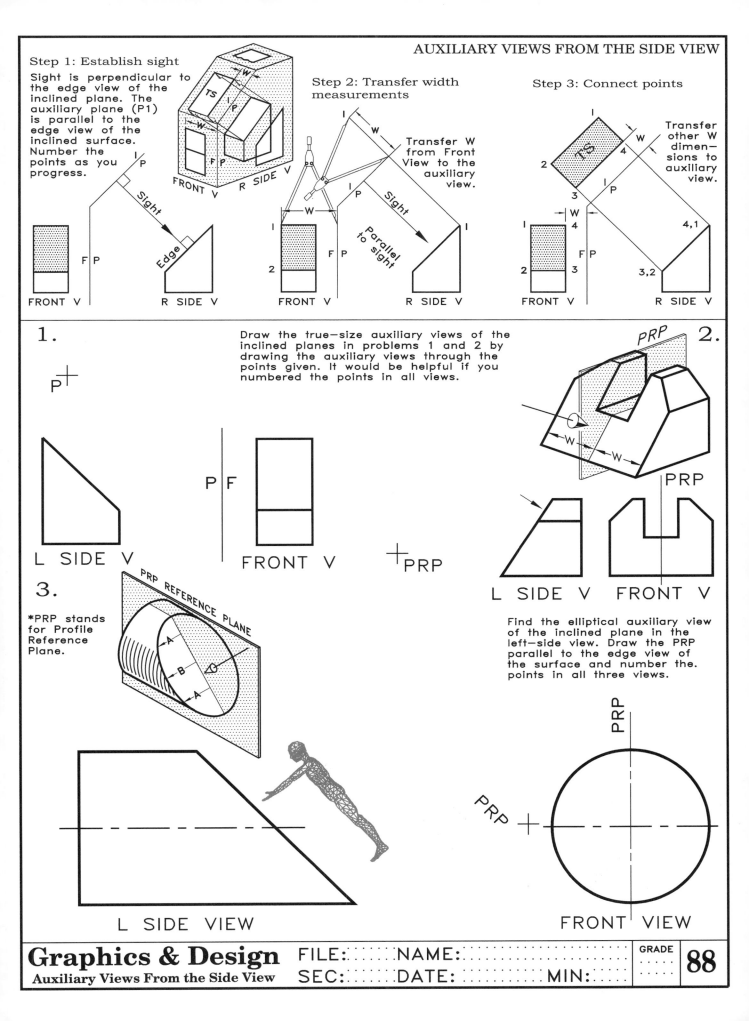

1.

Draw the true-size auxiliary views of the inclined planes in problems 1 and 2 by drawing the auxiliary views through the points given. It would be helpful if you numbered the points in all views.

2.

L SIDE V   FRONT V   PRP

Find the elliptical auxiliary view of the inclined plane in the left-side view. Draw the PRP parallel to the edge view of the surface and number the points in all three views.

3.

*PRP stands for Profile Reference Plane.

L SIDE VIEW

FRONT VIEW

Graphics & Design
Auxiliary Views From the Side View

FILE: .... NAME: ....
SEC: .... DATE: .... MIN: ....

GRADE **88**

6.4,6.4

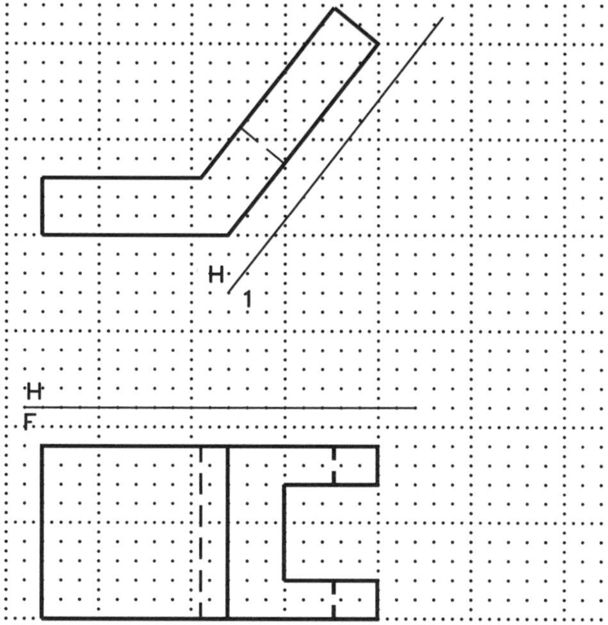

H 1

H
F

0,0

PRIMARY AUXILIARY VIEWS

DRAW THE GIVEN VIEWS AND PLOT THE
AUXILIARY VIEW OF THE INCLINED
SURFACE ONLY. ADD MISSING LINES AND
LABEL THE REFERENCE LINES.

Graphics & Design
Primary Auxiliary Views

FILE:........NAME:..................
SEC:.......DATE:..................

GRADE

88B

MIN:......

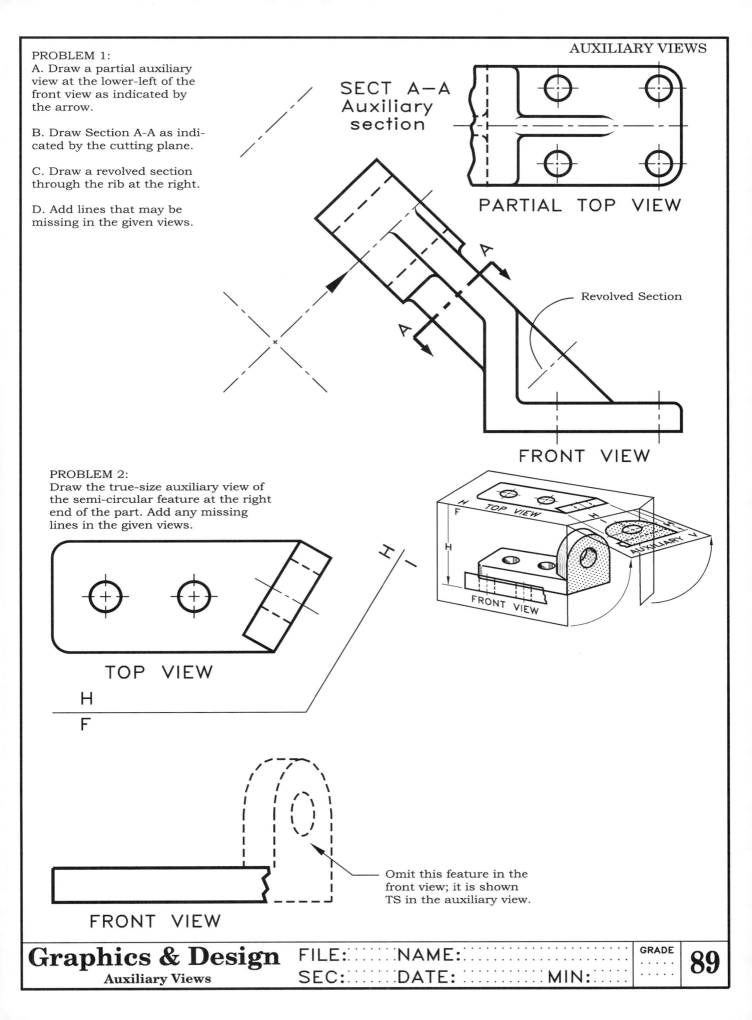

PROBLEM 1:

A. Draw a partial auxiliary view at the lower-left of the front view as indicated by the arrow.

B. Draw Section A-A as indicated by the cutting plane.

C. Draw a revolved section through the rib at the right.

D. Add lines that may be missing in the given views.

SECT A—A
Auxiliary
section

PARTIAL TOP VIEW

Revolved Section

FRONT VIEW

PROBLEM 2:
Draw the true-size auxiliary view of the semi-circular feature at the right end of the part. Add any missing lines in the given views.

TOP VIEW

TOP VIEW

FRONT VIEW

AUXILIARY V

H
F

H

Omit this feature in the front view; it is shown TS in the auxiliary view.

TOP VIEW

H
F

FRONT VIEW

Graphics & Design
Auxiliary Views

FILE: ...... NAME: ....
SEC: ...... DATE: .... MIN: ......

GRADE

89

6.4,6.4

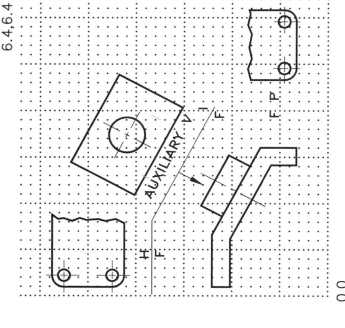

AUXILIARY V

J

F

H
F

F.P.

0,0

PRIMARY AUXILIARY VIEWS

COMPLETE THE GIVEN VIEWS, EACH OF
WHICH HAS MISSING LINES, JUST WAIT—
FOR YOUR ATTENTION. PROVIDE ANY
LABELING THAT YOU BELIEVE TO BE
NEEDED TO COMPLETE THE AUXILIARY
VIEW AND THE GIVEN VIEWS.

Graphics & Design
Primary Auxiliary Views

FILE: ...... NAME: ......

SEC: ...... DATE: ......

MIN: ......

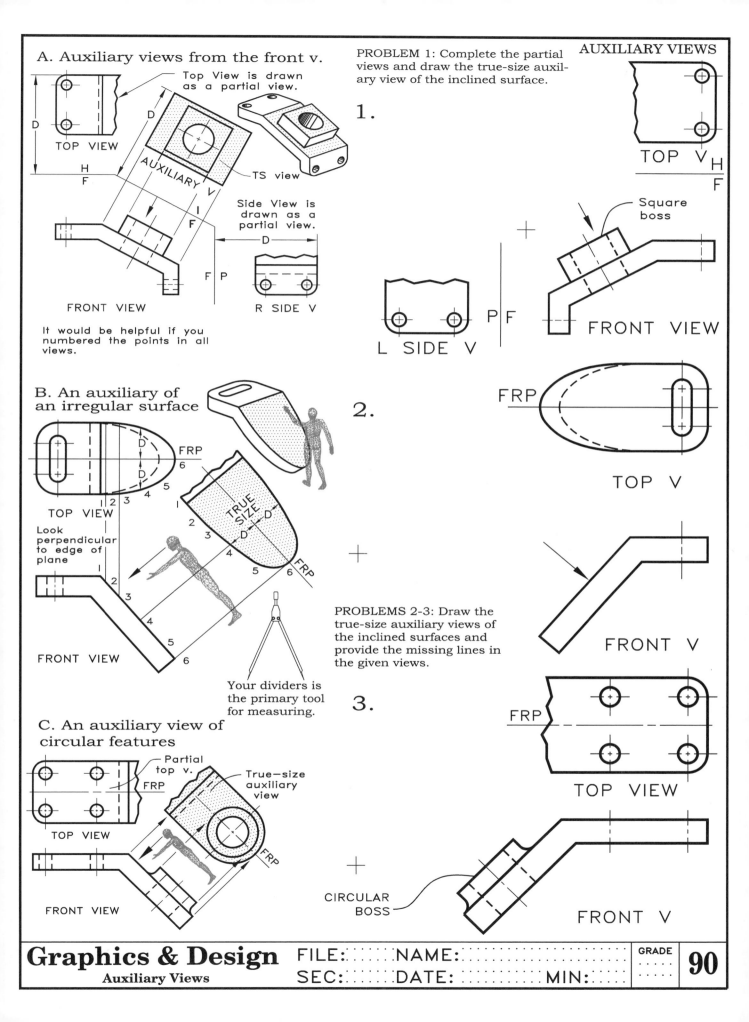

A. Auxiliary views from the front v.

Top View is drawn as a partial view.

TOP VIEW

AUXILIARY V

TS view

Side View is drawn as a partial view.

FRONT VIEW

R SIDE V

It would be helpful if you numbered the points in all views.

B. An auxiliary of an irregular surface

TOP VIEW

Look perpendicular to edge of plane

FRONT VIEW

TRUE SIZE

Your dividers is the primary tool for measuring.

C. An auxiliary view of circular features

Partial top v.

True−size auxiliary view

TOP VIEW

FRONT VIEW

PROBLEM 1: Complete the partial views and draw the true-size auxiliary view of the inclined surface.

AUXILIARY VIEWS

TOP V

1.

Square boss

L SIDE V

FRONT VIEW

FRP

TOP V

PROBLEMS 2-3: Draw the true-size auxiliary views of the inclined surfaces and provide the missing lines in the given views.

2.

FRONT V

3.

FRP

TOP VIEW

CIRCULAR BOSS

FRONT V

Graphics & Design
Auxiliary Views

FILE:⋯⋯ NAME:⋯⋯⋯⋯⋯⋯

SEC:⋯⋯ DATE:⋯⋯ MIN:⋯⋯

GRADE

90

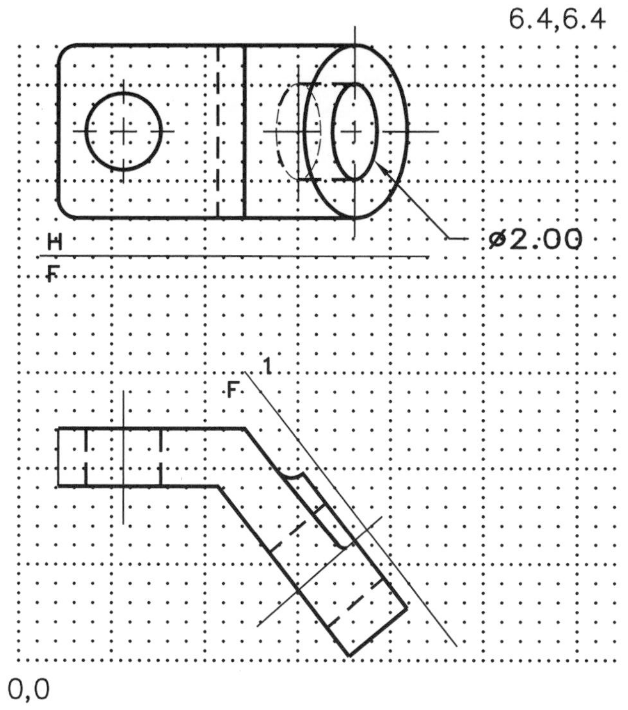

6.4,6.4

⌀2.00

H
F

F
1

0,0

PRIMARY AUXILIARY VIEWS

DRAW THE GIVEN VIEWS AND PLOT THE
AUXILIARY VIEW OF THE INCLINED
SURFACE ONLY. ADD MISSING LINES
AND LABEL THE REFERENCE LINES.

Graphics & Design
Auxiliary Views

FILE: .... NAME: ....
SEC: .... DATE: ....

GRADE

90B

MIN: ....

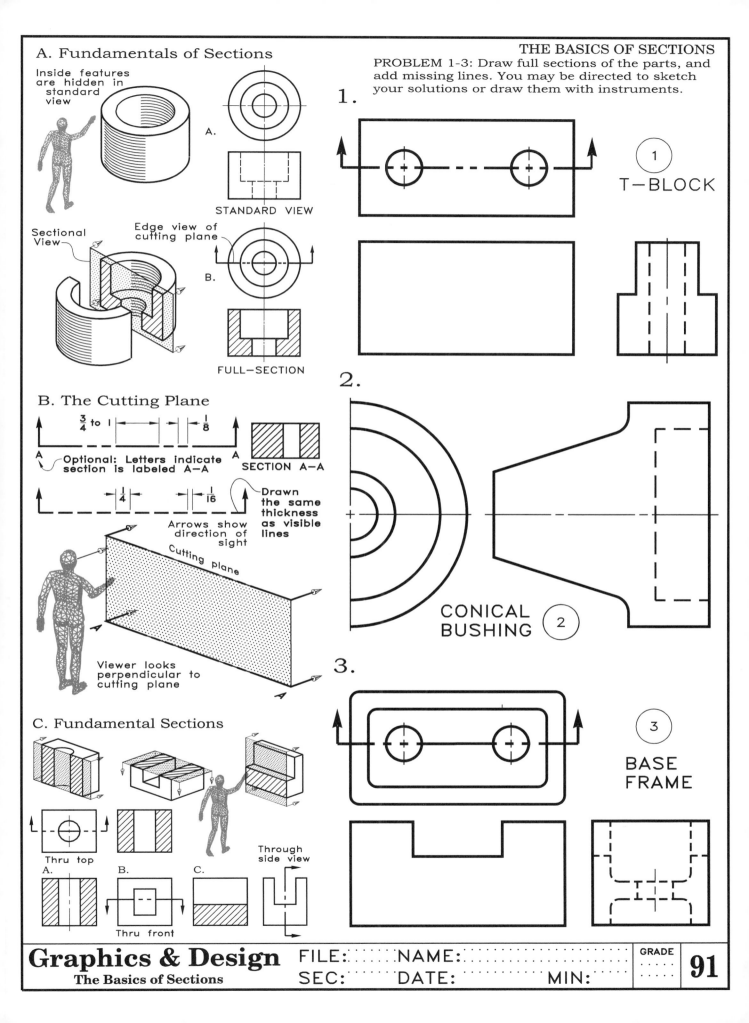

A. Fundamentals of Sections

Inside features are hidden in standard view

A.

STANDARD VIEW

Sectional View

Edge view of cutting plane

B.

FULL-SECTION

B. The Cutting Plane

$\frac{3}{4}$ to 1

$\frac{1}{8}$

Optional: Letters indicate section is labeled A–A

SECTION A–A

$\frac{1}{4}$

$\frac{1}{16}$

Drawn the same thickness as visible lines

Arrows show direction of sight

Cutting plane

Viewer looks perpendicular to cutting plane

C. Fundamental Sections

Thru top

A.

Thru front

B.

C.

Through side view

PROBLEM 1-3: Draw full sections of the parts, and add missing lines. You may be directed to sketch your solutions or draw them with instruments.

1.

① T–BLOCK

2.

CONICAL BUSHING ②

3.

③ BASE FRAME

2

6.4,6.4

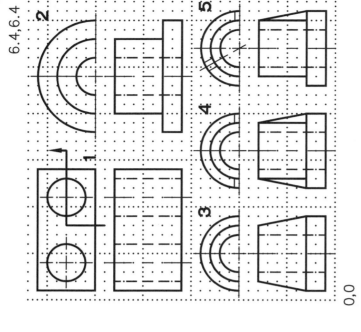

2.

1

5

4

3

0,0

HALF SECTIONS

PLOT THE GIVEN VIEWS OF THE PARTS
AND CONVERT THEM INTO HALF
SECTIONS. OMIT HIDDEN LINES WHERE
THEY ARE NOT NEEDED.

Graphics & Design
Half Sections

FILE: .... NAME: ....
SEC: .... DATE: ....

GRADE 91B
....
....
MIN:

A. Hatching Symbols for Sections

CAST IRON, MALLEABLE IRON		A GENERAL SYMBOL THAT CAN BE USE TO REPRESENT ALL MATERIALS.	
STEEL		ELECTRICAL WINDINGS, MAGNETS	
BRONZE, BRASS, COPPER		CONCRETE	
WHITE METAL, ZINC, LEAD, BABBITT		BRICK AND STONE MASONRY	
MAGNESIUM, ALUMINUM, AND ALLOYS		MARBLE, SLATE, GLASS PORCELAIN	
RUBBER, PLASTIC, ELECTRICAL INSUL.		EARTH	
CORK, FELT, FABRIC, FIBER, LEATHER		ROCK	

B. Crosshatching Techniques

1. GOOD
$\frac{1}{16} - \frac{1}{8}$ apart

2. POOR
Too close

3. POOR
Far apart

4. POOR
Uneven

5. POOR
Too heavy

6. POOR
Sloppy

C. Hatching Parts in Assembly

Outline section for large areas

Black–in thin parts

D. Hatching Multiple Parts

Section lines at varying angles for different parts

45°

45°

45°

30°

Same part, same angle

45°   45°   45°

1. THREE PARTS

2. TWO PARTS

PROBLEM 1: Draw crosshatching on the assembled sleeve and bushing.
PROBLEMS 2 & 3: Show cutting planes and the missing lines in each view in order for the front views to be full sections.

SLEEVE—CAST IRON

BUSHING—STEEL

1

2

THRU—3 HOLES

3

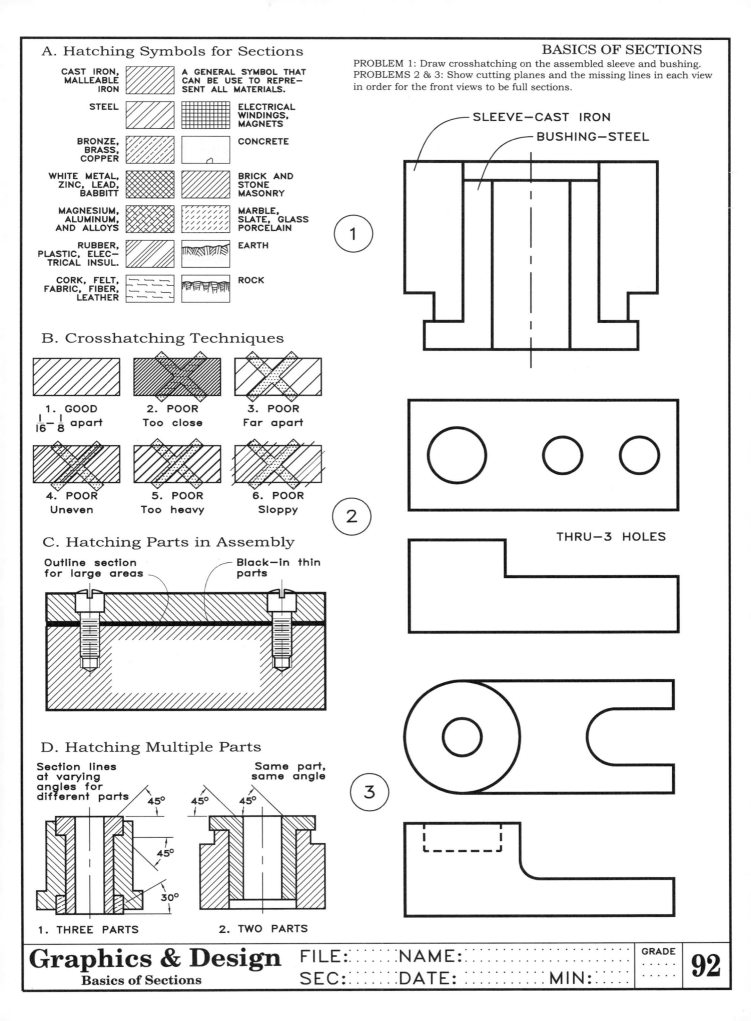

Graphics & Design
Basics of Sections

FILE: . . . . . . NAME: . . . .
SEC: . . . . . DATE: . . . . MIN: . . . . .

GRADE

92

Graphics & Design
Assignment Solution

FILE: .... NAME: ....

SEC: .... DATE: ....

GRADE ....

MIN: ....

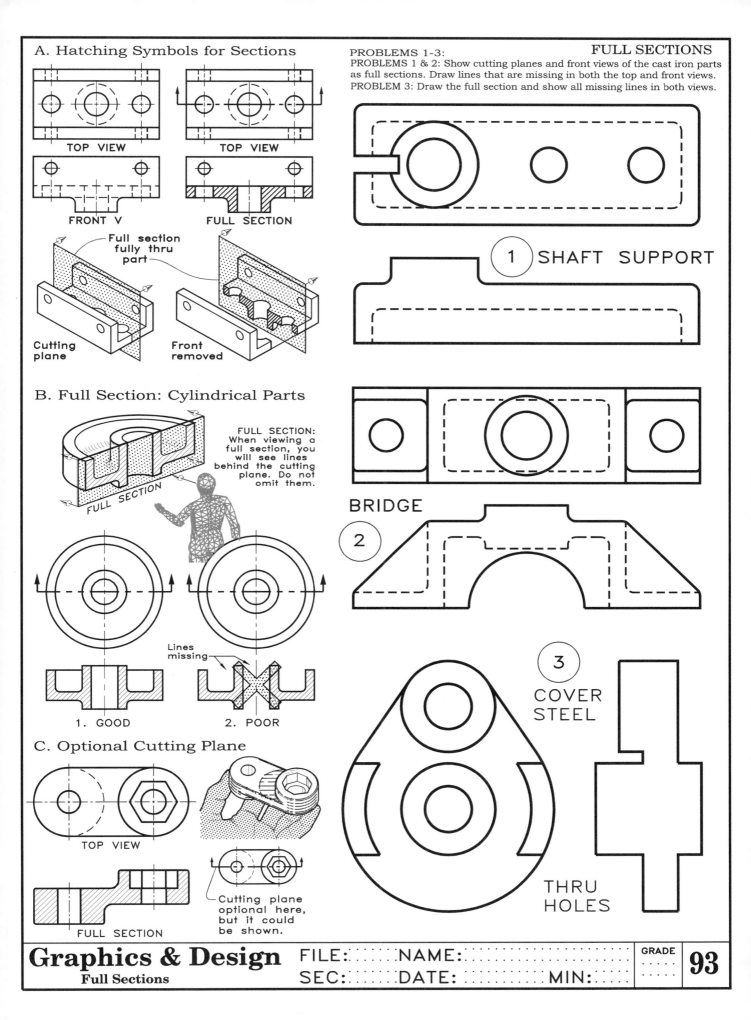

A. Hatching Symbols for Sections

TOP VIEW

TOP VIEW

FRONT V

FULL SECTION

Full section fully thru part

Cutting plane

Front removed

B. Full Section: Cylindrical Parts

FULL SECTION

FULL SECTION: When viewing a full section, you will see lines behind the cutting plane. Do not omit them.

Lines missing

1. GOOD

2. POOR

C. Optional Cutting Plane

TOP VIEW

FULL SECTION

Cutting plane optional here, but it could be shown.

PROBLEMS 1-3:
PROBLEMS 1 & 2: Show cutting planes and front views of the cast iron parts as full sections. Draw lines that are missing in both the top and front views.
PROBLEM 3: Draw the full section and show all missing lines in both views.

FULL SECTIONS

① SHAFT SUPPORT

BRIDGE

②

③ COVER STEEL

THRU HOLES

Graphics & Design
Full Sections

FILE: .... NAME: ....
SEC: ..... DATE: .... MIN: ....

GRADE

93

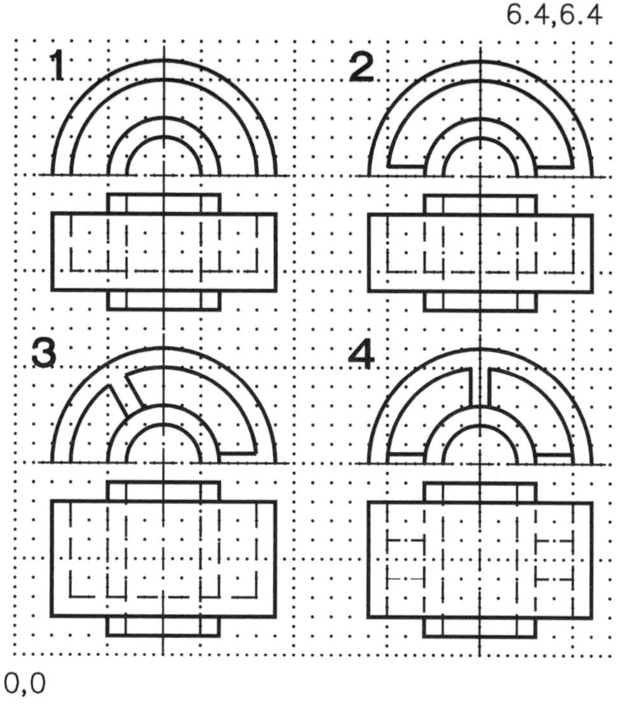

6.4,6.4

1

2

3

4

0,0

FULL SECTIONS

PLOT THE VIEWS OF THE PARTS AND
CONVERT THEM TO FULL SECTIONS.
OMIT HIDDEN LINES WHERE THEY ARE
NOT NEEDED.

Graphics & Design
Full Sections

FILE: .... NAME: ....
SEC: .... DATE: ....

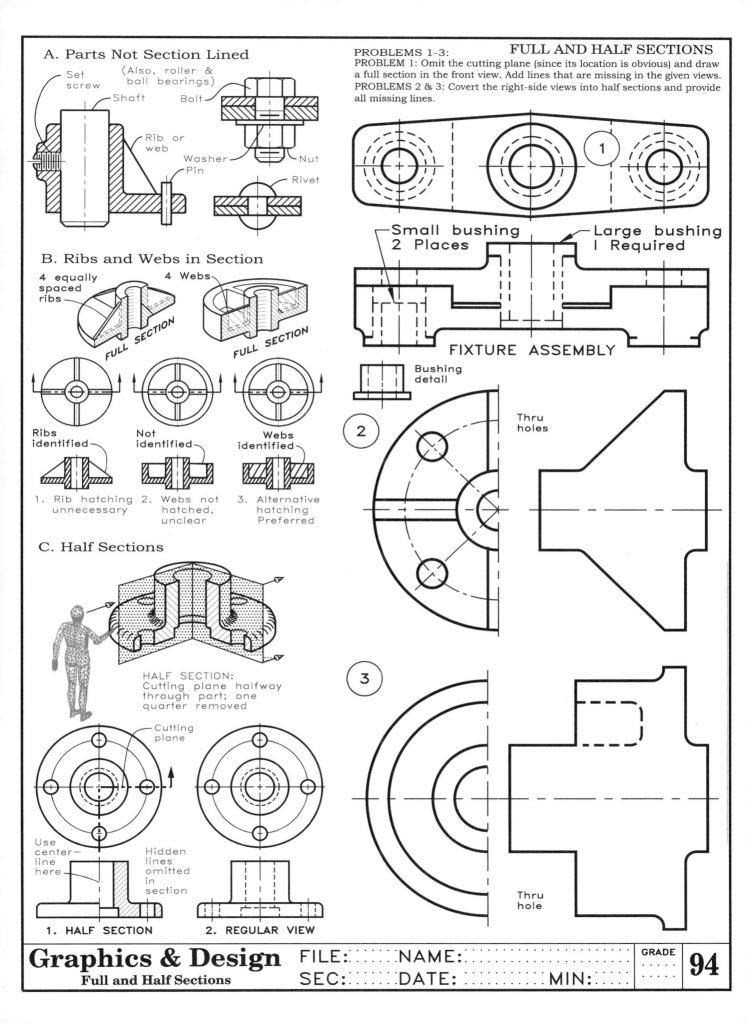

A. Parts Not Section Lined

(Also, roller & ball bearings)

Set screw
Shaft
Bolt
Rib or web
Washer
Pin
Nut
Rivet

B. Ribs and Webs in Section

4 equally spaced ribs
FULL SECTION

4 Webs
FULL SECTION

Ribs identified
Not identified
Webs identified

1. Rib hatching unnecessary
2. Webs not hatched, unclear
3. Alternative hatching Preferred

C. Half Sections

HALF SECTION: Cutting plane halfway through part; one quarter removed

Cutting plane

Use centerline here

Hidden lines omitted in section

1. HALF SECTION
2. REGULAR VIEW

PROBLEMS 1-3:   FULL AND HALF SECTIONS
PROBLEM 1: Omit the cutting plane (since its location is obvious) and draw a full section in the front view. Add lines that are missing in the given views.
PROBLEMS 2 & 3: Covert the right-side views into half sections and provide all missing lines.

(1)

Small bushing 2 Places
Large bushing 1 Required

FIXTURE ASSEMBLY

Bushing detail

(2)
Thru holes

(3)
Thru hole

Graphics & Design
Full and Half Sections

FILE: NAME:
SEC: DATE: MIN:
GRADE
94

Graphics & Design

Assignment Solution

FILE: NAME: : : : : : : : : : : : : : : : : :

SEC: DATE: : : : : : : : : : : : : : : : : :

MIN: : : : : : : : : : : : : : : : : :

GRADE

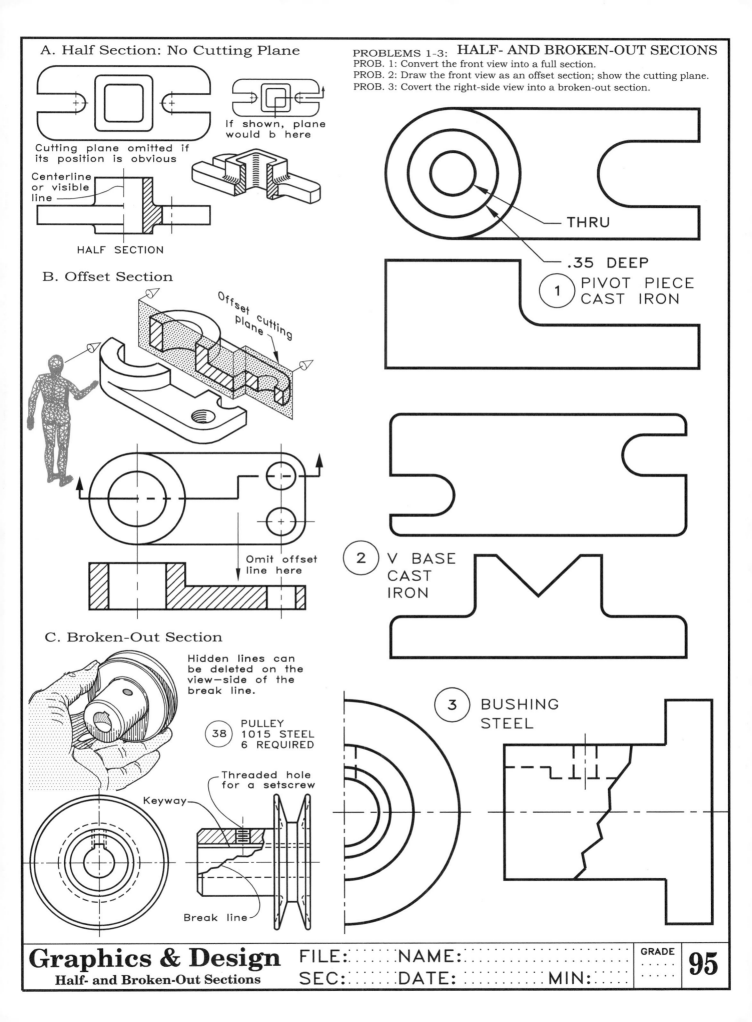

A. Half Section: No Cutting Plane

Cutting plane omitted if
its position is obvious

If shown, plane
would b here

Centerline
or visible
line

HALF SECTION

B. Offset Section

Offset cutting
plane

Omit offset
line here

C. Broken-Out Section

Hidden lines can
be deleted on the
view—side of the
break line.

(38) PULLEY
1015 STEEL
6 REQUIRED

Threaded hole
for a setscrew

Keyway

Break line

PROBLEMS 1-3: HALF- AND BROKEN-OUT SECIONS
PROB. 1: Convert the front view into a full section.
PROB. 2: Draw the front view as an offset section; show the cutting plane.
PROB. 3: Covert the right-side view into a broken-out section.

THRU

.35 DEEP

(1) PIVOT PIECE
CAST IRON

(2) V BASE
CAST
IRON

(3) BUSHING
STEEL

Graphics & Design
Half- and Broken-Out Sections

FILE:............ NAME:.................
SEC:........... DATE:............... MIN:.......

GRADE

95

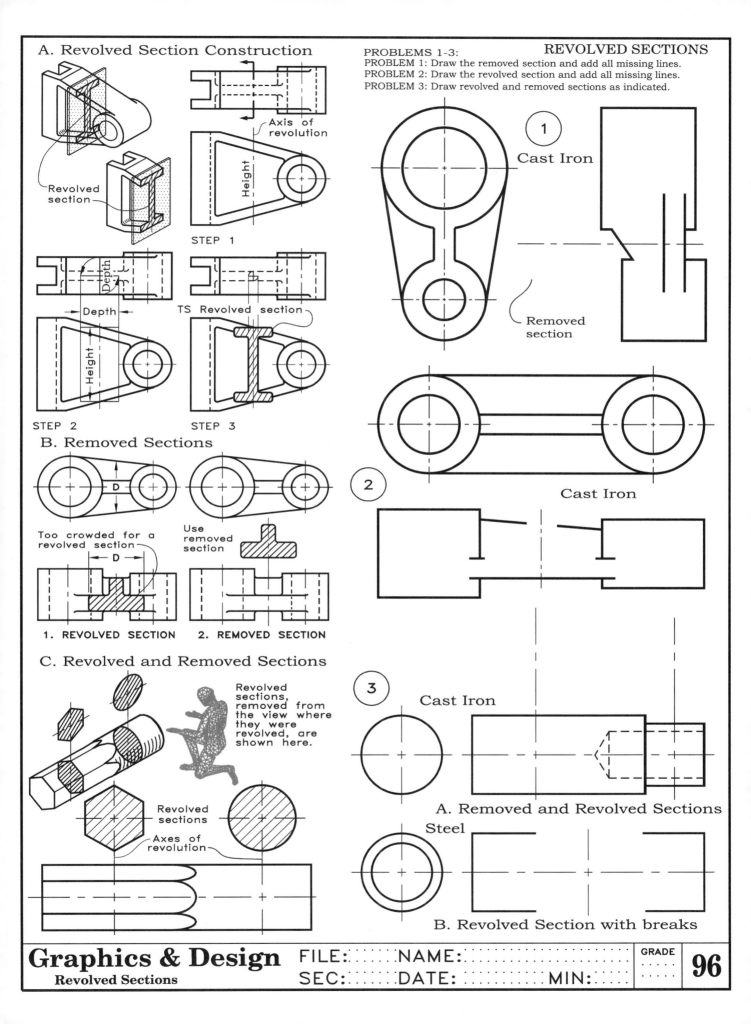

A. Revolved Section Construction

Revolved section

Axis of revolution

Height

STEP 1

Depth

Depth

Height

STEP 2

TS Revolved section

STEP 3

B. Removed Sections

D

Too crowded for a revolved section

Use removed section

D

1. REVOLVED SECTION

2. REMOVED SECTION

C. Revolved and Removed Sections

Revolved sections, removed from the view where they were revolved, are shown here.

Revolved sections

Axes of revolution

PROBLEMS 1-3:
PROBLEM 1: Draw the removed section and add all missing lines.
PROBLEM 2: Draw the revolved section and add all missing lines.
PROBLEM 3: Draw revolved and removed sections as indicated.

1

Cast Iron

Removed section

2

Cast Iron

3

Cast Iron

A. Removed and Revolved Sections

Steel

B. Revolved Section with breaks

Graphics & Design
Revolved Sections

FILE: .... NAME: ....

SEC: .... DATE: .... MIN: ....

GRADE

96

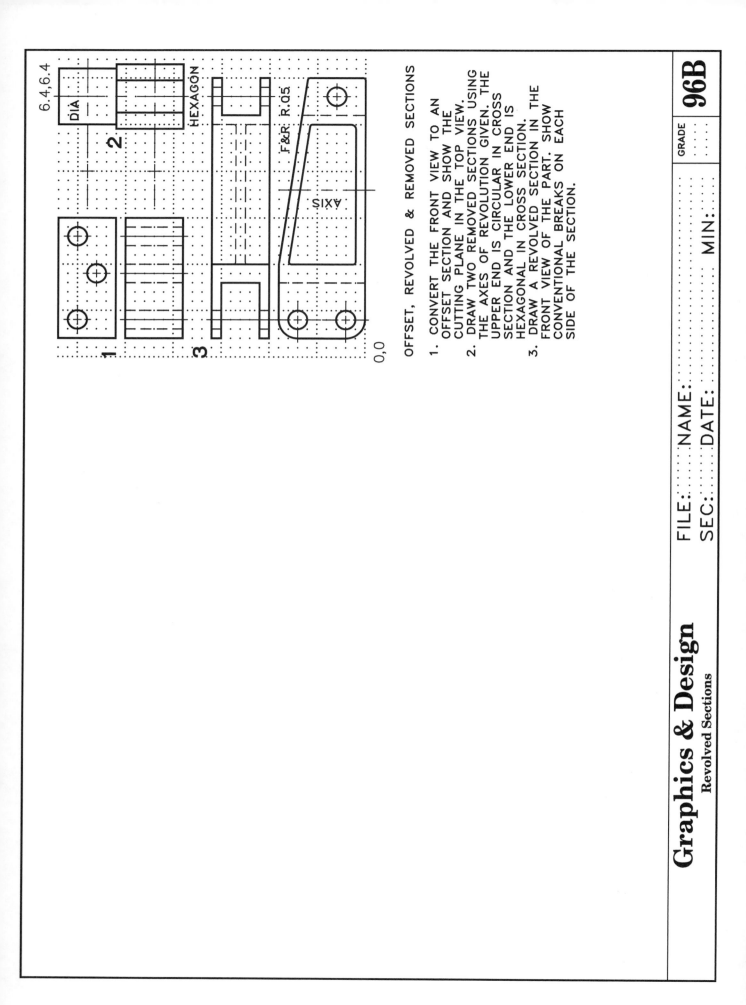

6.4,6.4

DIA

2.

HEXAGON

F&R R.05

AXIS

1.

3.

0,0

OFFSET, REVOLVED & REMOVED SECTIONS

1. CONVERT THE FRONT VIEW TO AN
   OFFSET SECTION AND SHOW THE
   CUTTING PLANE IN THE TOP VIEW.
2. DRAW TWO REMOVED SECTIONS USING
   THE AXES OF REVOLUTION GIVEN. THE
   UPPER END IS CIRCULAR IN CROSS
   SECTION AND THE LOWER END IS
   HEXAGONAL IN CROSS SECTION.
3. DRAW A REVOLVED SECTION IN THE
   FRONT VIEW OF THE PART. SHOW
   CONVENTIONAL BREAKS ON EACH
   SIDE OF THE SECTION.

Graphics & Design
Revolved Sections

FILE: NAME:

SEC: DATE:

A. Cylindrical and Tubular Breaks

$\frac{R}{3}$

— Freehand —

STEP 1   STEP 2   STEP 3

1. CYLINDRICAL BREAKS

$\frac{R}{2}$

— Freehand

STEP 1   STEP 2   STEP 3

2. TUBULAR BREAKS

B. Conventional-Break Symbols

RECTANGULAR (Wood)

RECTANGULAR (Metal)

TUBULAR (Metal)

CYLINDRICAL (Metal)

TUBULAR SECTION (Metal)

CYLINDRICAL SECTION (Metal)

RECTANGULAR: LONG BREAK (Any material)

C. Conventional-Break Options

4—2

Drawn true length

1.

Drawn less than TL

4—2

Drawn at a larger scale

2.

Conventional break

3.

Revolved Section

CONVENTIONAL PRACTICES

PROBLEMS 1-4:
PROBLEM 1: Draw a revolved section at the center with no conventional breaks.
PROBLEM 2: Draw a revolved section at the center with conventional breaks.
PROBLEM 3: Draw a revolved section in the front view with no breaks.
PROBLEM 4: Draw the revolved sections of the part as specified by the notes.

1

REVOLVED SECTION: WITHOUT BREAKS—C.I.

2

REVOLVED SECTION: WITH BREAKS—STEEL

3

REVOLVED SECTION: WITHOUT BREAKS

4

With breaks

Conventional breaks optional

Without conventional breaks

STEEL

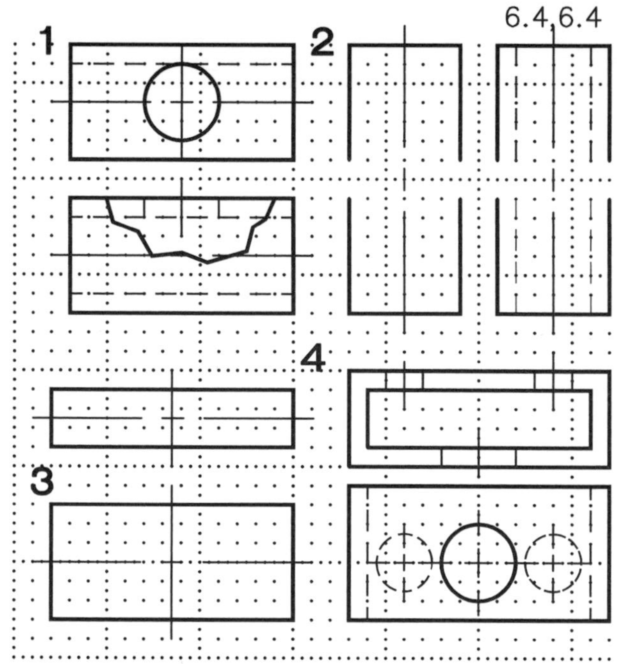

6.4,6.4

2.

1

4.

3

0,0

CONVENTIONAL PRACTICES

1. COMPLETE THE FRONT VIEW AS A
   BROKEN-OUT SECTION

2. DRAW CONVENTIONAL BREAKS FOR
   BOTH OF THE CYLINDRICAL PARTS.

3. DRAW REVOLVED SECTIONS WITHIN
   BOTH PARTS AND PLACE CONVEN-
   TIONAL BREAKS ON EACH SIDE
   OR OMIT THEM, AS ASSIGNED.

4. COMPLETE THE FRONT VIEW AS A
   PHANTOM SECTION.

Graphics & Design
Conventional Practices

FILE: . . . . . . NAME: . . . .

SEC: . . . . . . DATE: . . . .

GRADE **97B**

MIN: . . . .

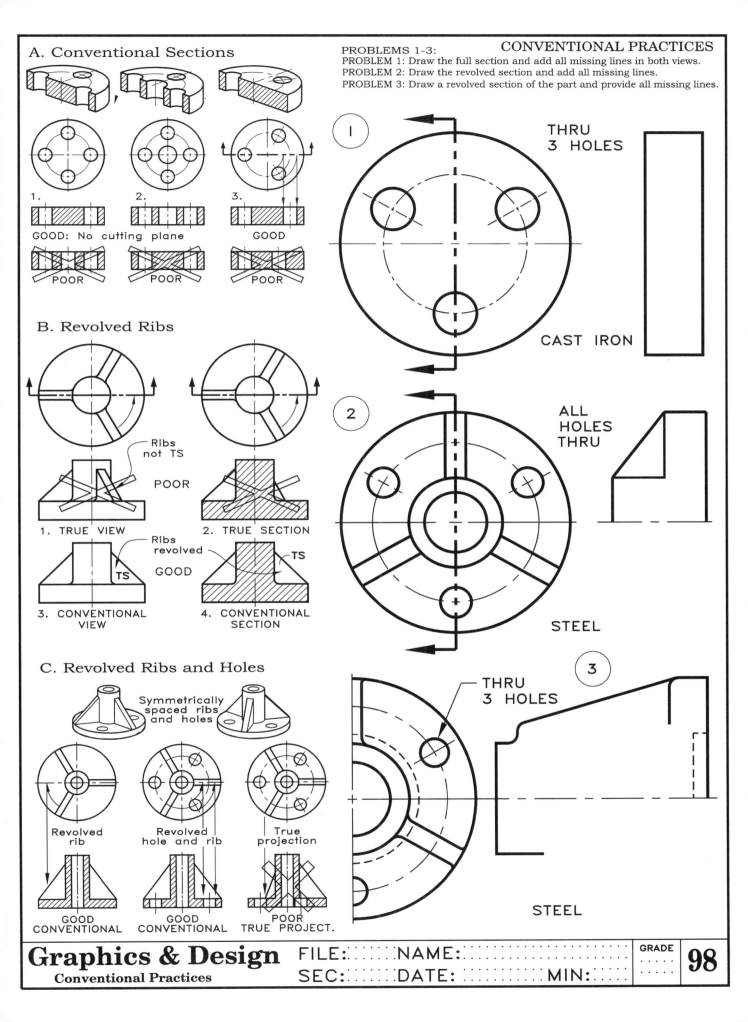

A. Conventional Sections

1. GOOD: No cutting plane
POOR

2. POOR

3. GOOD
POOR

B. Revolved Ribs

Ribs not TS
POOR

1. TRUE VIEW

2. TRUE SECTION

Ribs revolved
GOOD

TS

3. CONVENTIONAL VIEW

TS

4. CONVENTIONAL SECTION

C. Revolved Ribs and Holes

Symmetrically spaced ribs and holes

Revolved rib

Revolved hole and rib

True projection

GOOD CONVENTIONAL

GOOD CONVENTIONAL

POOR TRUE PROJECT.

PROBLEMS 1-3:
PROBLEM 1: Draw the full section and add all missing lines in both views.
PROBLEM 2: Draw the revolved section and add all missing lines.
PROBLEM 3: Draw a revolved section of the part and provide all missing lines.

1 THRU 3 HOLES

CAST IRON

2 ALL HOLES THRU

STEEL

3 THRU 3 HOLES

STEEL

Graphics & Design
Conventional Practices

FILE: .... NAME: ....

SEC: .... DATE: .... MIN: ......

GRADE

98

A. Rotated Lugs in Section

Lug

Draw lugs TS in Front Views & Sections

Rotate

1. VIEW

2. SECTION

Lugs are revolved to show their true size in the front view (1) and in the sectional view (2).

B. Spokes and Webs

Spokes

Solid web

Spokes

Do not hatch spokes

Solid web

1. FULL SECTION

2. FULL SECTION

Spokes are not hatched (1) so they are not confused with webs (2) that are hatched.

C. Rotation for Clarity

Draw front view as if the arm had been revolved into the frontal plane.

True location

Rotate to frontal plane

Do not hatch

True Shape

PROBLEMS 1-3:
PROBS. 1-2: Draw the full sections, rotate the features, and add the missing lines in each problem.
PROB. 3: Draw a revolved section of the part and provide all missing lines.

1

THRU
3 HOLES
CAST IRON

2

3 SPOKES
CAST IRON

3

THRU HOLES
3 FLANGES
STEEL

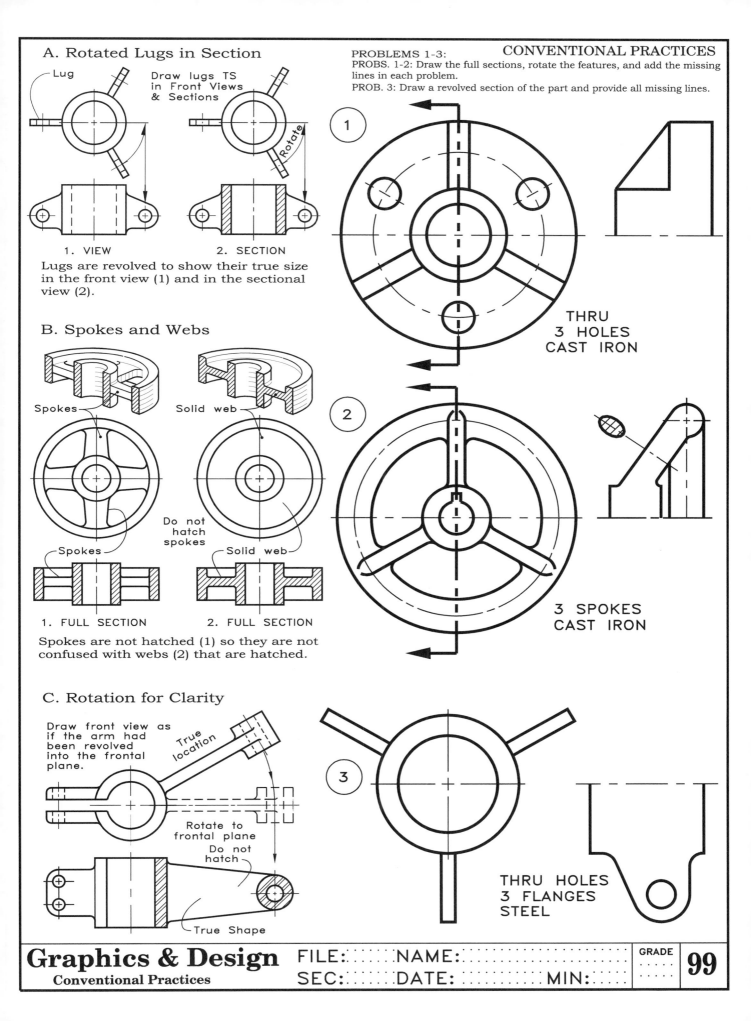

Graphics & Design
Conventional Practices

FILE: NAME:

SEC: DATE: MIN:

GRADE

99

A. Thread Terminology

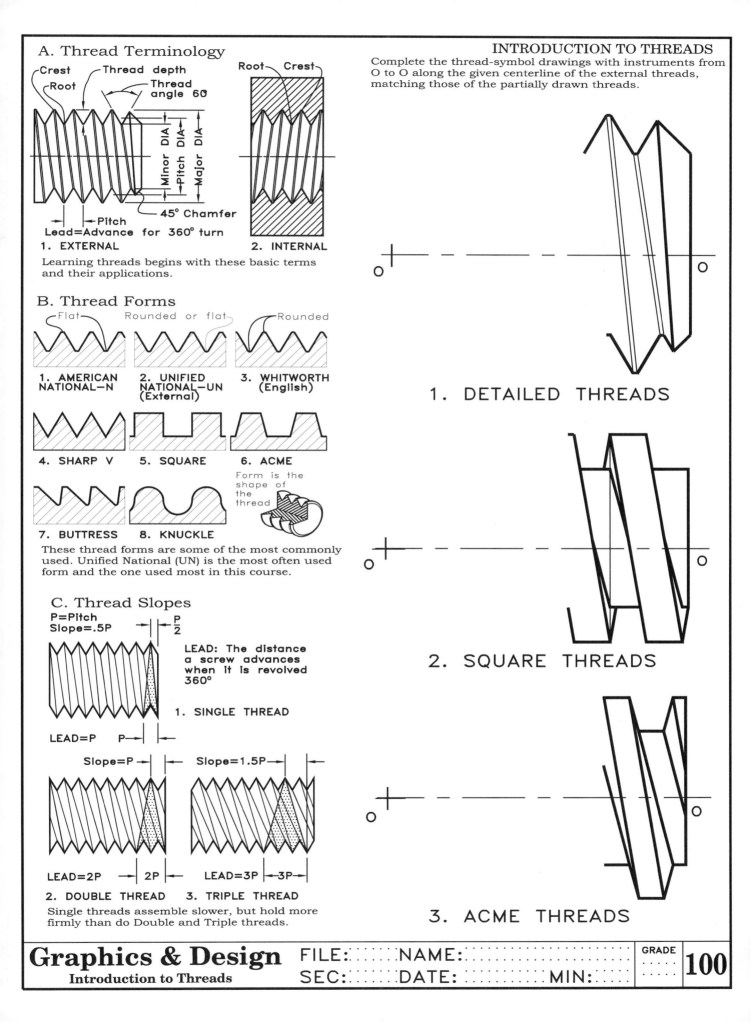

Crest
Root
Thread depth
Thread angle 60°
Minor DIA
Pitch DIA
Major DIA
45° Chamfer
Pitch
Lead=Advance for 360° turn

1. EXTERNAL

Root   Crest

2. INTERNAL

Learning threads begins with these basic terms and their applications.

B. Thread Forms

Flat

1. AMERICAN NATIONAL—N

Rounded or flat

2. UNIFIED NATIONAL—UN (External)

Rounded

3. WHITWORTH (English)

4. SHARP V

5. SQUARE

6. ACME

7. BUTTRESS

8. KNUCKLE

Form is the shape of the thread

These thread forms are some of the most commonly used. Unified National (UN) is the most often used form and the one used most in this course.

C. Thread Slopes

P=Pitch
Slope=.5P
$\frac{P}{2}$

LEAD: The distance a screw advances when it is revolved 360°

1. SINGLE THREAD

LEAD=P   P

Slope=P

Slope=1.5P

LEAD=2P   2P

LEAD=3P   3P

2. DOUBLE THREAD   3. TRIPLE THREAD

Single threads assemble slower, but hold more firmly than do Double and Triple threads.

Complete the thread-symbol drawings with instruments from O to O along the given centerline of the external threads, matching those of the partially drawn threads.

O   O

1. DETAILED THREADS

O   O

2. SQUARE THREADS

O   O

3. ACME THREADS

Graphics & Design
Introduction to Threads

FILE: NAME:
SEC: DATE: MIN:

GRADE **100**

A. Basic Thread Notes

Reference: Page 102 has data American threads.

Reference: Page 102 has data American threads.

Major diameter
Threads per inch
Form (Unified Natl.)
Series (Coarse)
Class of fit (Med.)
A=external; B=Int.
LH=left hand
No note for RH

—1.50—6UNC—3B

Arrow to the visible circle

1. INTERNAL

—1.00—8UNC—2A—LH

Add **DOUBLE** if a double thread

2. EXTERNAL

B. Typical Thread Notes

UNR form applies to external threads only

2.50—4UNRC—2A

Or
.50—13UNC—2A
½ —13UNC—2A

Notes as decimal or common fractions

Or
10—32UNF—2A
.19—32UNF—2A

1. UNR THREADS

2. THREAD NOTES

C. American National Standards Institute Unified Inch Threads (UN and UNR)

Nominal Diameter	Basic Diameter	Coarse NC & UNC		Fine NF & UNF		Extra Fine NEF/UNEF	
		Thds per inch	Tap Drill Dia	Thds per inch	Tap Drill Dia	Thds per inch	Tap Drill Dia
1	1.000	8	.875	12	.922	20	.953
1–1/16	1.063	...	...	...	...	18	1.000
1–1/8	1.125	7	.904	12	1.046	18	1.070
1–3/16	1.188	...	...	...	...	18	1.141
1–1/4	1.250	7	1.109	12	1.172	18	1.188
1–5/16	1.313	...	...	...	...	18	1.266
1–3/8	1.375	6	1.219	12	1.297	18	1.313
1–7/16	1.438	...	...	...	...	18	1.375
1–1/2	1.500	6	1.344	12	1.422	18	1.438

D. Types of Thead Symbols

P

1. DETAILED
For medium or large parts

2. SCHEMATIC
For medium or large parts
P

3. SIMPLIFIED
For small parts

Thread pitches are drawn larger than true size for better symbols.

P= Pitch
P

THREAD SPECIFICATIONS: ENGLISH

PROBLEMS 1-10: Apply thread notes to these external threads using the information given below and the table at C.

1. DIA=1.50 UNC FIT=2 LH   2. DIA=1.00 UN   FIT=2 RH
3. DIA=1.25 UNF FIT=2 LH   4. DIA=1.00 UNC FIT=2 RH

1. SCHEMATIC

2. SIMPLIFIED

3. DETAILED

4. SCHEMATIC: BOLT

5. DIA=1.00 NEF FIT=2 LH   6. DIA=1.00 NF FIT=3 RH
7. DIA=1.25 UNC FIT=2 RH   8. DIA=1.25 UN FIT=1 RH
9. DIA=1.50 NEF FIT=2 RH   10. DIA=1.50 NC FIT=2 RH

END VIEW   THRU
5. SIMPLIFIED

END VIEW   THRU
6. SECTION

END VIEW   THRU
7. SCHEMATIC VIEW

END VIEW   THRU
8. SECTION

END VIEW   THRU
9. DETAILED VIEW

END VIEW   THRU
10. SECTION

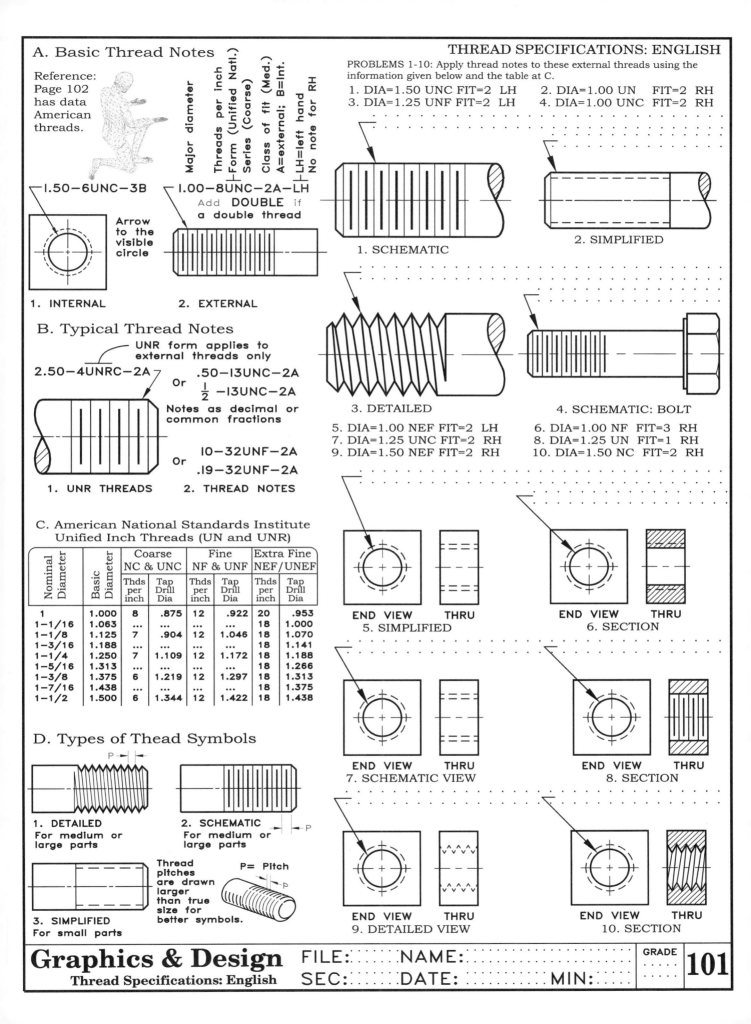

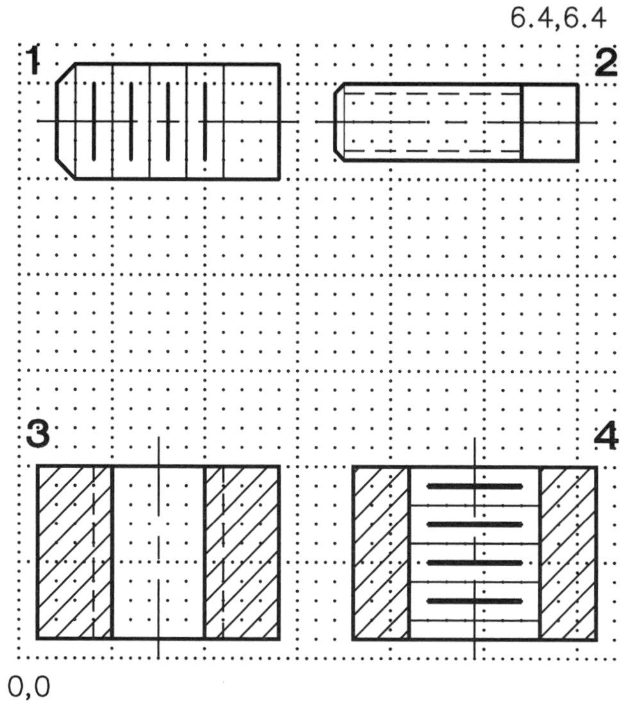

6.4,6.4

2

4

1

3

0,0

THREAD SPECIFICATIONS: ENGLISH

DRAW THE THREADED PARTS AND
ADD THE FOLLOWING NOTES:

1. DIA=1.25; COARSE, CLASS 2 FIT

2. DIA=.625; COARSE; CLASS 2 FIT

3. DIA=1.50; FINE; CLASS 1 FIT

4. DIA=1.375; EXTRA FINE; CLASS 2

Graphics & Design
Thread Specifications: Englisih

FILE: .... NAME: ....
SEC: .... DATE: ....

GRADE
101B
....
MIN: ....

APPENDIX 1: METRIC SCREW THREADS: AMERICAN NATIONAL AND UNIFIED (Inches)

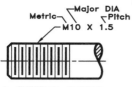

A. EXTERNAL THREAD

NOTE: TAP–DRILL DIAMETER IS APPROXIMATELY 75% OF THE MAJOR DIAMETER.

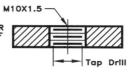

B. INTERNAL THREAD

COARSE		FINE		COARSE		FINE	
MAJ. DIA & THD PITCH	TAP DRILL	MAJ. DIA & THD PITCH	TAP DRILL	MAJ. DIA & THD PITCH	TAP DRILL	MAJ. DIA & THD PITCH	TAP DRILL
M1.6 X 0.35	1.25			M20 X 2.5	17.5	M20 X 1.5	18.5
M1.8 X 0.35	1.45			M22 X 2.5	19.5	M22 X 1.5	20.5
M2 X 0.4	1.6			M24 X 3	21.0	M24 X 2	22.0
M2.2 X 0.45	1.75			M27 X 3	24.0	M27 X 2	25.0
M2.5 X 0.45	2.05			M30 X 3.5	26.5	M30 X 2	28.0
M3 X 0.5	2.5			M33 X 3.5	29.5	M33 X 2	31.0
M3.5 X 0.6	2.9			M36 X 4	32.0	M36 X 3	33.0
M4 X 0.7	3.3			M39 X 4	35.0	M39 X 3	36.0
M4.5 X 0.75	3.75			M42 X 4.5	37.5	M42 X 3	39.0
M5 X 0.8	4.2			M45 X 4.5	40.5	M45 X 3	42.0
M6 X 1	5.0			M48 X 5	43.0	M48 X 3	45.0
M7 X 1	6.0			M52 X 5	47.0	M52 X 3	49.0
M8 X 1.25	6.8	M8 X 1	7.0	M56 X 5.5	50.5	M56 X 4	52.0
M9 X 1.25	7.75			M60 X 5.5	54.5	M60 X 4	56.0
M10 X 1.5	8.5	M10 X 1.25	8.75	M64 X 6	58.0	M64 X 4	60.0
M11 X 1.5	9.5			M68 X 6	62.0	M68 X 4	64.0
M12 X 1.75	10.3	M12 X 1.25	10.5	M72 X 6	66.0	M72 X 4	68.0
M14 X 2	12.0	M14 X 1.5	12.5	M80 X 6	74.0	M80 X 4	76.0
M16 X 2	14.0	M16 X 1.5	14.5	M90 X 6	84.0	M90 X 4	86.0
M18 X 2.5	15.5	M18 X 1.5	16.5	M100 X 6	94.0	M100 X 4	96.0

ANSI/AME B1.1

APPENDIX 2: SQUARE THREADS AND ACME THREADS

2.00–2.5 SQUARE

A typical thread note

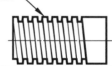

Dimensions are in inches

Size	Size	Thds per inch	Size	Size	Thds per inch	Size	Size	Thds per inch
3/8	.375	12	1-1/8	1.125	4	3	3.000	1-1/2
7/16	.438	10	1-1/4	1.250	4	3-1/4	3.125	1-1/2
1/2	.500	10	1-1/2	1.500	3	3-1/2	3.500	1-1/3
9/16	.563	8	1-3/4	1.750	2-1/2	3-3/4	3.750	1-1/3
5/8	.625	8	2	2.000	2-1/2	4	4.000	1-1/3
3/4	.75	6	2-1/4	2.250	2	4-1/4	4.250	1-1/3
7/8	.875	5	2-1/2	2.500	2	4-1/2	4.500	1
1	1.000	5	2-3/4	2.750	2	Larger		1

APPENDIX 3: AMER. STANDARD TAPER PIPE THREADS (NPT)

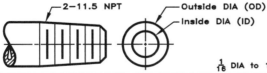

2–11.5 NPT — Outside DIA (OD) — Inside DIA (ID)

PIPES THRU 12 INCHES IN DIA ARE SPECIFIED BY THEIR INSIDE DIAMETERS. LARGER PIPES ARE SPECIFIED BY THEIR OD,

$\frac{1}{16}$ DIA to $1\frac{1}{4}$ DIA   Dimensions in inches

Nominal ID	$\frac{1}{16}$	$\frac{1}{8}$	$\frac{1}{4}$	$\frac{3}{8}$	$\frac{1}{2}$	$\frac{3}{4}$	1	$1\frac{1}{4}$
Outisde DIA	0.313	0.405	0.540	0.675	0.840	1.050	1.315	1.660
Thds/Inch	27	27	18	18	14	14	$11\frac{1}{2}$	$11\frac{1}{2}$

$1\frac{1}{2}$ DIA to 6 DIA

Nominal ID	$1\frac{1}{2}$	2	$2\frac{1}{2}$	3	$3\frac{1}{2}$	4	5	6
Outisde DIA	1.900	2.375	2.875	3.500	4.000	4.500	5.563	6.625
Thds/Inch	$11\frac{1}{2}$	$11\frac{1}{2}$	8	8	8	8	8	8

8 DIA to 24 DIA

Nominal ID	8	10	12	14 OD	16 OD	18 OD	20 OD	24 OD
Outisde DIA	8.625	10.750	12.750	14.000	16.000	18.000	20.000	24.000
Thds/Inch	8	8	8	8	8	8	8	8

SOURCE: ANSI B2.1

APPENDIX 4
HEXAGON-HEAD BOLTS

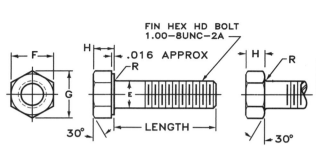

FIN HEX HD BOLT
1.00—8UNC—2A

.016 APPROX

LENGTH

30°   30°

*10 MEANS THAT LENGTHS ARE AVAILABLE
AT 1 INCH INCREMENTS UP TO 10 INCHES.

Dimensions are in inches

DIA	E Max.	F Max.	G Avg.	H Max.	R Max.	DIA	E Max.	F Max.	G Avg.	H Max.	R Max.
1/4	.250	.438	.505	.163	.025	1—1/8	1.125	1.688	1.949	.718	.095
5/16	.313	.500	.577	.211	.025	1—1/4	1.250	1.875	2.165	.813	.095
3/8	.375	.563	.650	.243	.025	1—3/8	1.375	2.063	2.382	.878	.095
7/16	.438	.625	.722	.291	.025	1—1/2	1.500	2.250	2.598	.974	.095
1/2	.500	.750	.866	.323	.025	1—3/4	1.750	2.625	3.031	1.134	.095
9/16	.563	.812	.938	.371	.045	2	2.000	3.000	3.464	1.263	.095
5/8	.625	.938	1.083	.403	.045	2—1/4	2.250	3.375	3.897	1.423	.095
3/4	.750	1.125	1.299	.483	.045	2—1/2	2.500	3.750	4.330	1.583	.095
7/8	.875	1.313	1.516	.563	.065	2—3/4	2.750	4.125	4.763	1.744	.095
1	1.000	1.500	1.732	.627	.095	3	3.000	4.500	5.196	1.935	.095

APPENDIX 5
HEX NUTS AND HEX JAM NUTS

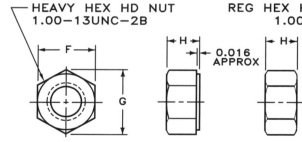

HEAVY HEX HD NUT
1.00—13UNC—2B

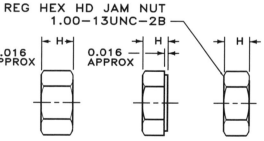

REG HEX HD JAM NUT
1.00—13UNC—2B

0.016 APPROX   0.016 APPROX

HEAVY HEX NUTS
AND HEX JAM NUTS

REGULAR HEX   HEX JAM
NUT   NUT

MAJOR DIA		F Max.	G Avg.	H Max.	H Max.	MAJOR DIA		F Max.	G Avg.	H Max.	H Max.
1/4	.250	.438	.505	.226	.163	3/4	.750	1.125	1.299	.665	.446
5/16	.313	.500	.577	.273	.195	7/8	.875	1.313	1.516	.776	.510
3/8	.375	.563	.650	.337	.227	1	1.000	1.500	1.732	.887	.575
7/16	.438	.688	.794	.385	.260	1—1/8	1.125	1.688	1.949	.899	.639
1/2	.500	.750	.866	.448	.323	1—1/4	1.250	1.875	2.165	1.094	.751
9/16	.563	.875	1.010	.496	.324	1—3/8	1.375	2.063	2.382	1.206	.815
5/8	.625	.938	1.083	.559	.387	1—1/2	1.500	2.250	2.598	1.317	.880

APPENDIX
AP102B

A. Metric Thread Notes

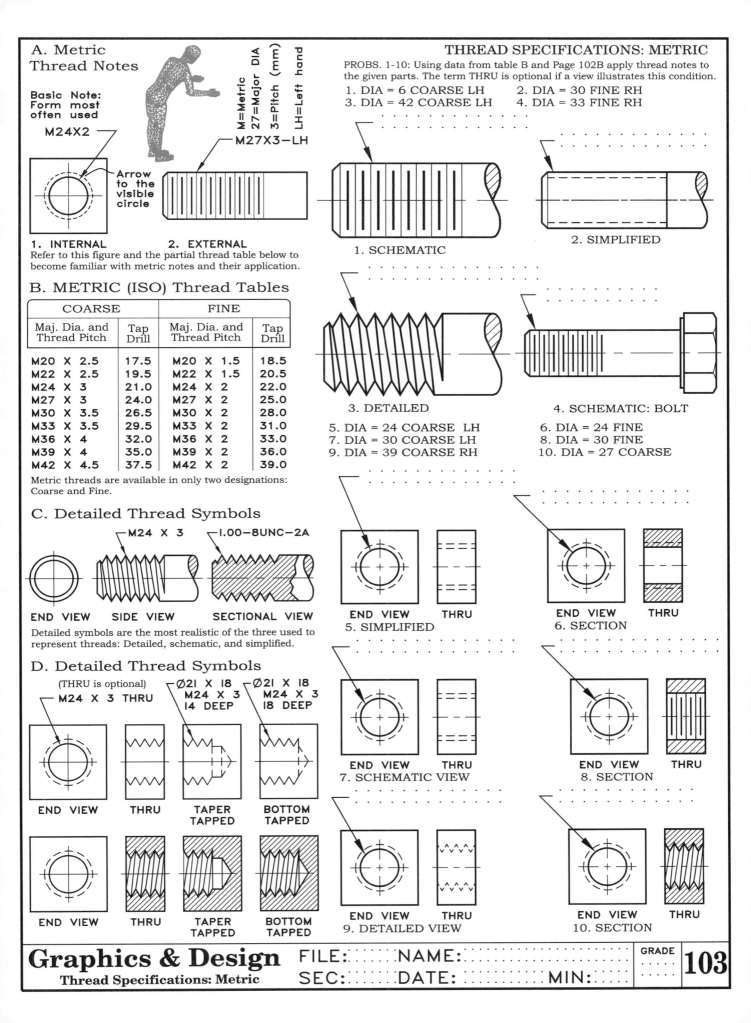

Basic Note:
Form most
often used

M24X2

M=Metric
27=Major DIA
3=Pitch (mm)
LH=Left hand

M27X3—LH

Arrow to the visible circle

1. INTERNAL   **2. EXTERNAL**

Refer to this figure and the partial thread table below to become familiar with metric notes and their application.

B. METRIC (ISO) Thread Tables

COARSE		FINE	
Maj. Dia. and Thread Pitch	Tap Drill	Maj. Dia. and Thread Pitch	Tap Drill
M20 X 2.5	17.5	M20 X 1.5	18.5
M22 X 2.5	19.5	M22 X 1.5	20.5
M24 X 3	21.0	M24 X 2	22.0
M27 X 3	24.0	M27 X 2	25.0
M30 X 3.5	26.5	M30 X 2	28.0
M33 X 3.5	29.5	M33 X 2	31.0
M36 X 4	32.0	M36 X 2	33.0
M39 X 4	35.0	M39 X 2	36.0
M42 X 4.5	37.5	M42 X 2	39.0

Metric threads are available in only two designations: Coarse and Fine.

C. Detailed Thread Symbols

—M24 X 3   —1.00—8UNC—2A

END VIEW   SIDE VIEW   SECTIONAL VIEW

Detailed symbols are the most realistic of the three used to represent threads: Detailed, schematic, and simplified.

D. Detailed Thread Symbols

(THRU is optional)
—M24 X 3 THRU

Ø21 X 18
M24 X 3
14 DEEP

Ø21 X 18
M24 X 3
18 DEEP

END VIEW   THRU   TAPER TAPPED   BOTTOM TAPPED

END VIEW   THRU   TAPER TAPPED   BOTTOM TAPPED

THREAD SPECIFICATIONS: METRIC

PROBS. 1-10: Using data from table B and Page 102B apply thread notes to the given parts. The term THRU is optional if a view illustrates this condition.

1. DIA = 6 COARSE LH   2. DIA = 30 FINE RH
3. DIA = 42 COARSE LH   4. DIA = 33 FINE RH

1. SCHEMATIC   **2. SIMPLIFIED**

3. DETAILED   **4. SCHEMATIC: BOLT**

5. DIA = 24 COARSE LH   6. DIA = 24 FINE
7. DIA = 30 COARSE LH   8. DIA = 30 FINE
9. DIA = 39 COARSE RH   10. DIA = 27 COARSE

END VIEW   THRU
5. SIMPLIFIED

END VIEW   THRU
6. SECTION

END VIEW   THRU
7. SCHEMATIC VIEW

END VIEW   THRU
8. SECTION

END VIEW   THRU
9. DETAILED VIEW

END VIEW   THRU
10. SECTION

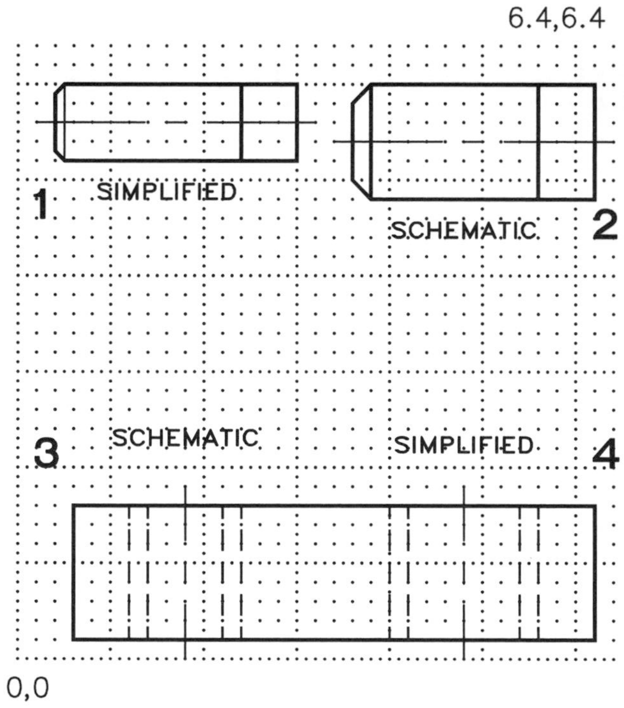

6.4,6.4

2

SCHEMATIC

4

SIMPLIFIED

1

SIMPLIFIED

3

SCHEMATIC

0,0

METRIC THREADS

PLOT THE VIEWS OF THE THREADED
PARTS SHOWN ABOVE, GIVE THEIR
THREAD SYMBOLS, AND GIVE THEIR
THREAD NOTES IN ACCORDANCE WITH
THE SPECIFICATIONS BELOW:

1. DIA=20, COARSE

2. DIA=30, COARSE

3. DIA=20, FINE

4. DIA=39, COARSE

Graphics & Design
Metric Threads

FILE: .... NAME: ....

SEC: .... DATE: ....

GRADE 103B
MIN:

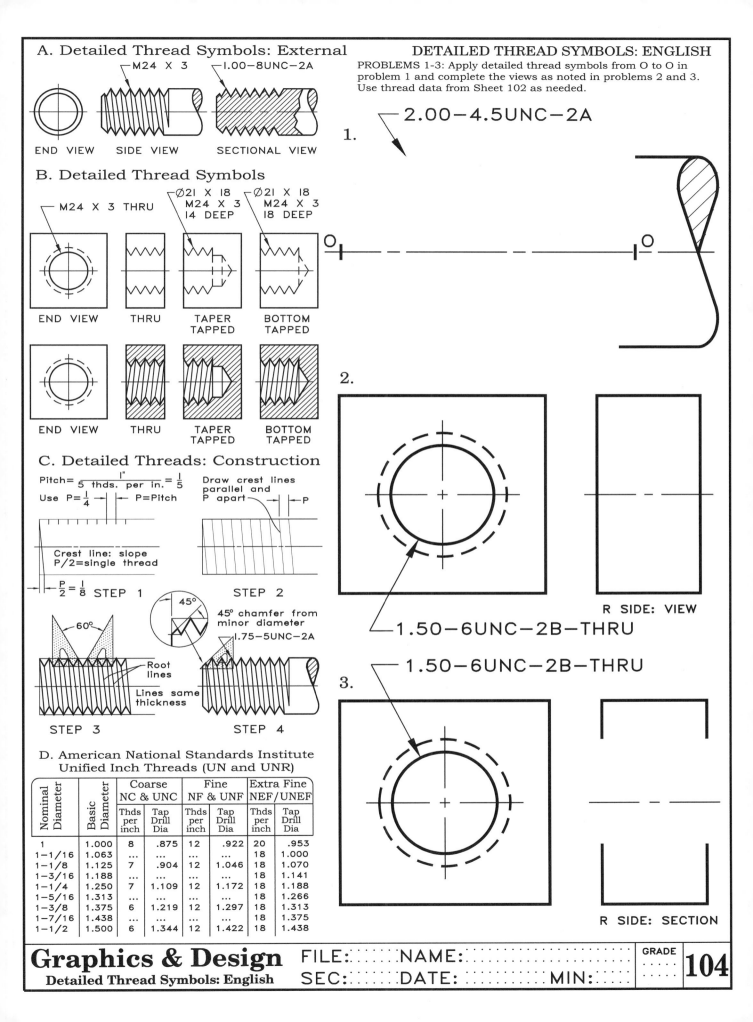

A. Detailed Thread Symbols: External

M24 X 3   1.00—8UNC—2A

END VIEW   SIDE VIEW   SECTIONAL VIEW

B. Detailed Thread Symbols

M24 X 3 THRU

Ø21 X 18
M24 X 3
14 DEEP

Ø21 X 18
M24 X 3
18 DEEP

END VIEW   THRU   TAPER TAPPED   BOTTOM TAPPED

END VIEW   THRU   TAPER TAPPED   BOTTOM TAPPED

C. Detailed Threads: Construction

Pitch $= \dfrac{1''}{5 \text{ thds. per in.}} = \dfrac{1}{5}$

Use $P = \dfrac{1}{4}$ P=Pitch

Draw crest lines parallel and P apart

Crest line: slope P/2=single thread

$\dfrac{P}{2} = \dfrac{1}{8}$ STEP 1

STEP 2

60°

Root lines

Lines same thickness

45°

45° chamfer from minor diameter
1.75—5UNC—2A

STEP 3   STEP 4

D. American National Standards Institute Unified Inch Threads (UN and UNR)

Nominal Diameter	Basic Diameter	Coarse NC & UNC		Fine NF & UNF		Extra Fine NEF/UNEF	
		Thds per inch	Tap Drill Dia	Thds per inch	Tap Drill Dia	Thds per inch	Tap Drill Dia
1	1.000	8	.875	12	.922	20	.953
1–1/16	1.063	...	...	...	...	18	1.000
1–1/8	1.125	7	.904	12	1.046	18	1.070
1–3/16	1.188	...	...	...	...	18	1.141
1–1/4	1.250	7	1.109	12	1.172	18	1.188
1–5/16	1.313	...	...	...	...	18	1.266
1–3/8	1.375	6	1.219	12	1.297	18	1.313
1–7/16	1.438	...	...	...	...	18	1.375
1–1/2	1.500	6	1.344	12	1.422	18	1.438

PROBLEMS 1-3: Apply detailed thread symbols from O to O in problem 1 and complete the views as noted in problems 2 and 3. Use thread data from Sheet 102 as needed.

1.   2.00—4.5UNC—2A

O   O

2.

R SIDE: VIEW

1.50—6UNC—2B—THRU

1.50—6UNC—2B—THRU

3.

R SIDE: SECTION

Graphics & Design
Detailed Thread Symbols: English

FILE: NAME:

SEC: DATE: MIN:

GRADE

104

A. A Properly Noted Bolt

.50-13UNC-2A X 2.25 LONG
REG HEX HEAD CAP SCREW

1.25
Length

PROBLEM 1: Draw the hexagon head of a bolt at the right of the half view by referring to the four-step example at B and the given thread note.

B. Drawing a Hexagon Head

DIA
1½ DIA

Dimensions are based on the major DIA of the bolt

⅔ DIA

STEP 1

Across corners

Washer face approx. 1/16

STEP 2

60°
R

STEP 3

30° Chamfer tangent to arc

STEP 4

1.

FINISHED HEX HD BOLT
1.50-6UNC-2A

PROBLEM 2: Draw the hexagon head of a bolt below the half view by referring to the four-step example at the left and the given thread note.

2.

C. Drawing a Square Head

DIA
1½ DIA

Dimensions are based on the major DIA of the bolt

⅔ DIA

STEP 1

Across Corners

STEP 2

60°
R   R

STEP 3

30° Chamfer tangent to arc

STEP 4

SQUARE HEAD BOLT
1.375-6UNC-2A

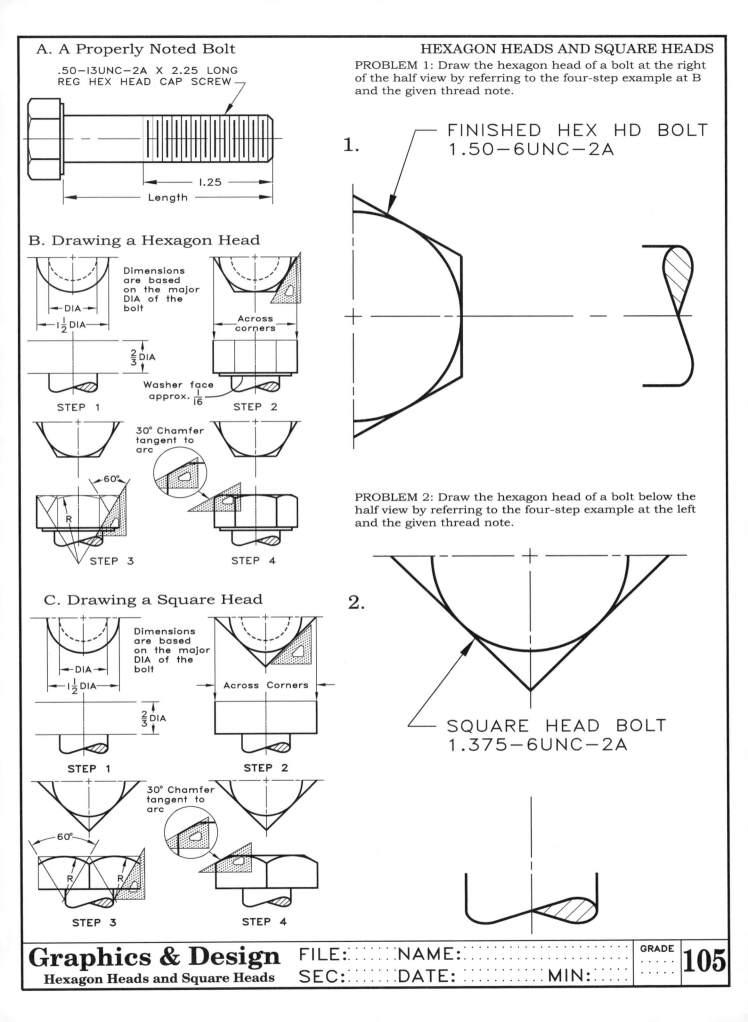

Graphics & Design
Hexagon Heads and Square Heads

FILE: NAME:
SEC: DATE: MIN:

GRADE
105

Graphics & Design
Assignment Solution

FILE: :::::::::: NAME: :::
SEC: :::::::::: DATE: :::

GRADE
:::: ::::
:::: ::::

A. Schematic Symbols: External

M24X3 M24X3

END VIEW SIDE VIEW SECTION

B. Schematic Symbols: Internal

M24 X 3

Ø21 X 18
M24X3
14 DEEP

Ø21 X 18
M24X3
18 DEEP

END VIEW THRU TAPER TAPPED BOTTOM TAPPED

END VIEW THRU TAPER TAPPED BOTTOM TAPPED

C. Drawing Schematic Threads

Min. DIA is approx. 75% of major DIA

Chamfer 45° from minor DIA

$\frac{3}{4}$ DIA

Minor DIA

STEP 1 STEP 2

True pitch= $\frac{1}{10}$

Use $\frac{1}{8}$ pitch

P= $\frac{1}{8}$

Thick root lines

Thread note .75-10UNC-2A

Thin crest lines

STEP 3 STEP 4

*See Sheet 102 for the complete tables.

D. American National Standards Institute Unified Inch Threads (UN and UNR)

Nominal Diameter	Basic Diameter	Coarse NC & UNC		Fine NF & UNF		Extra Fine NEF/UNEF	
		Thds per inch	Tap Drill Dia	Thds per inch	Tap Drill Dia	Thds per inch	Tap Drill Dia
1	1.000	8	.875	12	.922	20	.953
1-1/16	1.063	...	...	...	...	18	1.000
1-1/8	1.125	7	.904	12	1.046	18	1.070
1-3/16	1.188	...	...	...	...	18	1.141
1-1/4	1.250	7	1.109	12	1.172	18	1.188
1-5/16	1.313	...	...	...	...	18	1.266
1-3/8	1.375	6	1.219	12	1.297	18	1.313
1-7/16	1.438	...	...	...	...	18	1.375
1-1/2	1.500	6	1.344	12	1.422	18	1.438

SCHEMATIC THREAD SYMBOLS: ENGLISH

PROBLEMS 1-4: Apply schematic thread symbols to O and thread notes from the table at D. Follow the examples at the left.
1. Left end: DIA= 1.50, 2A fit, 6 threads per inch, LH
2. Right end: DIA= 1.50, 2A fit, 6 threads per inch
3. DIA= 1.00, 2A fit, 8 threads per inch
4. DIA= 1.125, 2A fit, 12 threads per inch

1. 2.

STUD

3. 4.

FILLISTER HEAD HEXAGON HEAD

PROBLEMS 5-6: Complete the internal view and section and add notes to each by referring to the examples and the ANSI table.
5. DIA=1.00 NC, Drill 1.12 Deep, Thread .92 deep.
6. DIA=1.00 NF, Drill 1.00 Deep, Thread 1.00 deep.

5.

Thread to here
Drill to here

FRONT VIEW R SIDE: VIEW

6.

Thread to here

FRONT VIEW R SIDE: SECTION

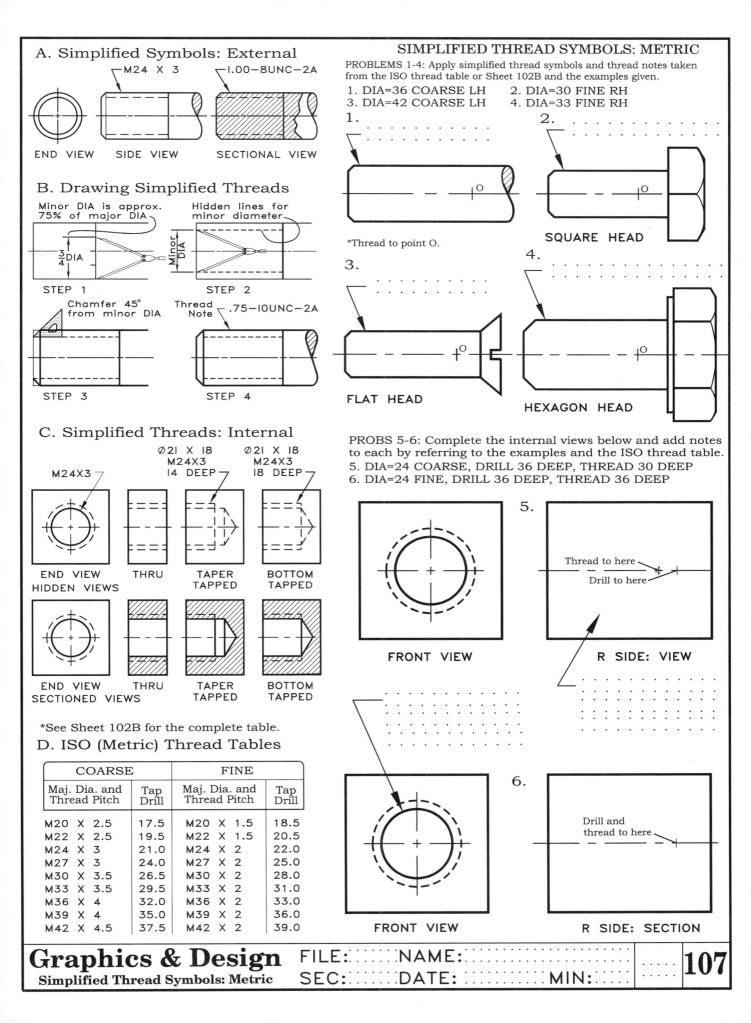

A. Simplified Symbols: External

—M24 X 3   —1.00–8UNC–2A

END VIEW   SIDE VIEW   SECTIONAL VIEW

B. Drawing Simplified Threads

Minor DIA is approx. 75% of major DIA.

Hidden lines for minor diameter.

¾ DIA   Minor DIA

STEP 1   STEP 2

Chamfer 45° from minor DIA

Thread Note   .75–10UNC–2A

STEP 3   STEP 4

C. Simplified Threads: Internal

M24X3

Ø21 X 18   Ø21 X 18
M24X3   M24X3
14 DEEP   18 DEEP

END VIEW   THRU   TAPER   BOTTOM
HIDDEN VIEWS   TAPPED   TAPPED

END VIEW   THRU   TAPER   BOTTOM
SECTIONED VIEWS   TAPPED   TAPPED

*See Sheet 102B for the complete table.

D. ISO (Metric) Thread Tables

COARSE		FINE	
Maj. Dia. and Thread Pitch	Tap Drill	Maj. Dia. and Thread Pitch	Tap Drill
M20 X 2.5	17.5	M20 X 1.5	18.5
M22 X 2.5	19.5	M22 X 1.5	20.5
M24 X 3	21.0	M24 X 2	22.0
M27 X 3	24.0	M27 X 2	25.0
M30 X 3.5	26.5	M30 X 2	28.0
M33 X 3.5	29.5	M33 X 2	31.0
M36 X 4	32.0	M36 X 2	33.0
M39 X 4	35.0	M39 X 2	36.0
M42 X 4.5	37.5	M42 X 2	39.0

SIMPLIFIED THREAD SYMBOLS: METRIC

PROBLEMS 1-4: Apply simplified thread symbols and thread notes taken from the ISO thread table or Sheet 102B and the examples given.

1. DIA=36 COARSE LH   2. DIA=30 FINE RH
3. DIA=42 COARSE LH   4. DIA=33 FINE RH

1.

2.

*Thread to point O.

SQUARE HEAD

3.

4.

FLAT HEAD

HEXAGON HEAD

PROBS 5-6: Complete the internal views below and add notes to each by referring to the examples and the ISO thread table.
5. DIA=24 COARSE, DRILL 36 DEEP, THREAD 30 DEEP
6. DIA=24 FINE, DRILL 36 DEEP, THREAD 36 DEEP

5.

Thread to here
Drill to here

FRONT VIEW   R SIDE: VIEW

6.

Drill and thread to here

FRONT VIEW   R SIDE: SECTION

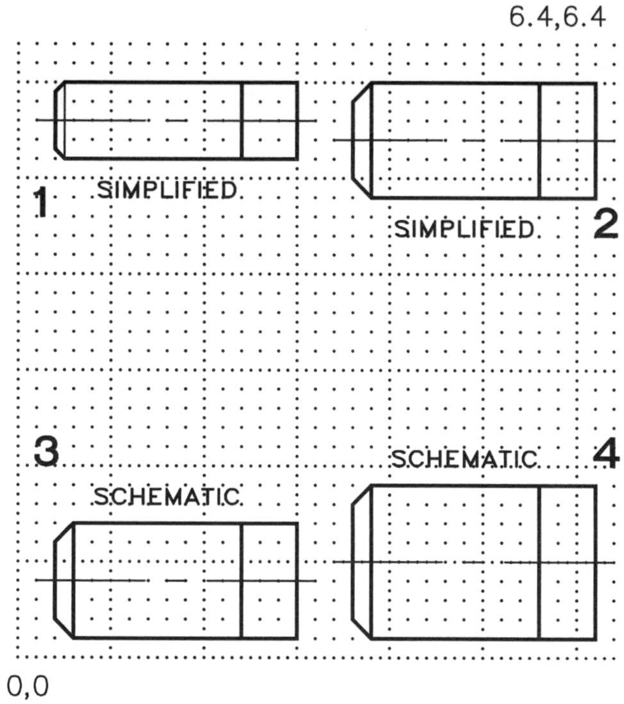

6.4,6.4

1 SIMPLIFIED.

2 SIMPLIFIED.

3 SCHEMATIC.

4 SCHEMATIC.

0,0

THREAD NOTES: ENGLISH

PLOT THE VIEWS OF THE THREADED
PARTS SHOWN ABOVE, GIVE THEIR
THREAD SYMBOLS, AND GIVE THEIR
THREAD NOTES IN ACCORDANCE WITH
THE SPECIFICATIONS BELOW:

1. DIA=.75, COARSE, FIT=2

2. DIA=1.25, FINE, FIT=3

3. DIA=1.25, COARSE, FIT=2

4. DIA=1.625, EXTRA FINE, FIT=2

Graphics & Design
Thread Notes: English

FILE: NAME:
SEC: DATE:

GRADE 107B
MIN:

A. Geometry of Nuts and Bolts

$\frac{3}{8}$–16UNC–2AX1 HEX HD BOLT & NUT

60° R D D D D 1$\frac{1}{2}$D

Across corners $\frac{2}{3}$D Length of thds $\frac{7}{8}$D Across flats

$\frac{5}{8}$–11UNC–2AX1 SQ HD BOLT & NUT

60° R D 2D 1$\frac{1}{2}$D

$\frac{2}{3}$D LENGTH $\frac{7}{8}$D

B. Examples of Cap Screws

1.00–8UNC–2A X 2.00 HEX HD CAP SCREW

1.00–8UNC–2A X 2.20 FLAT HD CAP SCR

1. HEX HEAD   2. FLAT HEAD

1–8UNC–2A X 3 ROUND HD CAP SCREW

1.00–8UNC–2A X 3–FILLISTER HD CAP SCR

1–8UNC–2A HEX SOC HD CAP SCR

3. ROUND HD   4. FILLISTER HD   5. HEX SOCKET

C. Examples of Machine Screws

M14X2X34–FILLISTER HD MACH SCREW

M14X1.5X30 FLAT HEAD SCREW

1. FILLISTER HEAD   2. FLAT HEAD

M12X1.25X30–OVAL HD MACHINE SCREW

M12X1.75X30–ROUND HD MACH SCREW

3. OVAL HEAD   4. ROUND HEAD

These screws are drawn on a grid to give proportions for drawing them at different sizes. Notes are used to provide thread specifications and other information.

TYPES OF NUTS AND BOLTS

PROBLEMS 1-5: Draw schematic thread symbols for the partially-drawn nuts and bolts in accordance with their thread notes. Add missing lines in all views and draw the end views of problems 4 and 5.

1. .75–12UN–2AX1.5 HEX HD BOLT & NUT

Thread to here

2. .75–12UNC–2AX1.5 SQUARE HD BOLT & NUT

Thread to here

UNC bolt - Square head

3. 1.125–7UNC–2AX2.5 FLAT HD CAP SCREW

Flat head cap screw

4. 1.50–6UNC–2AX2.5 HEX SOCKET HD CAP SCREW

Hexagon socket head screw

5. 1.00–8UNC–2AX2.5 OVAL HD MACH. SCREW

Oval head machine screw

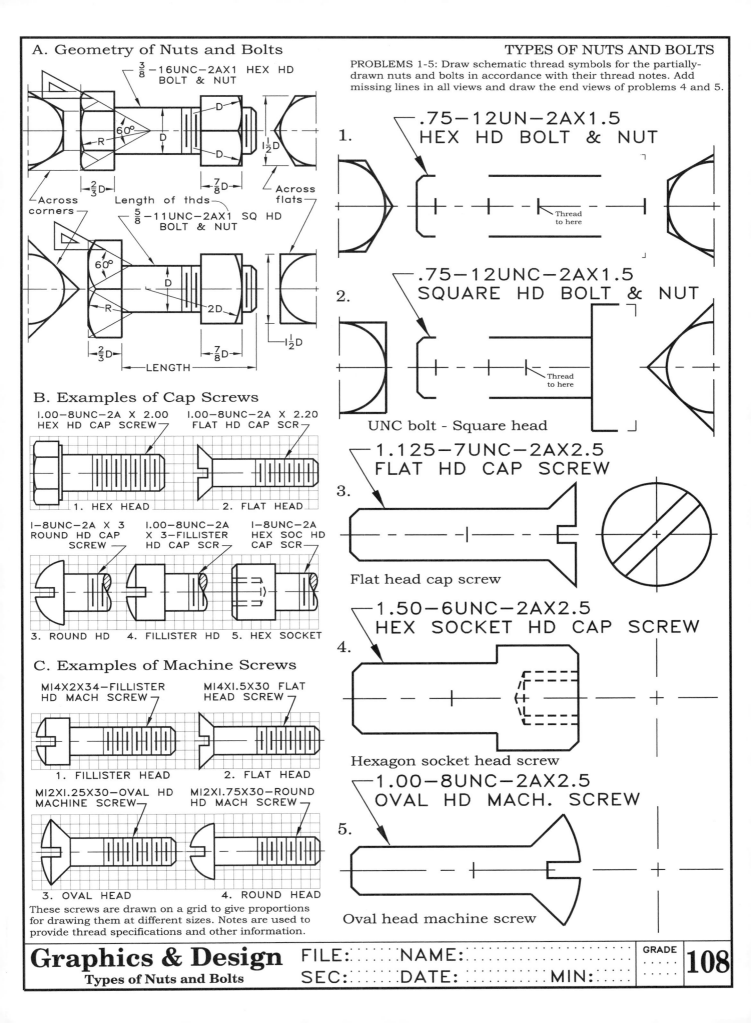

Graphics & Design
Types of Nuts and Bolts

FILE: .... NAME: ....
SEC: .... DATE: .... MIN: ....

GRADE

108

A. Examples of Setscrews

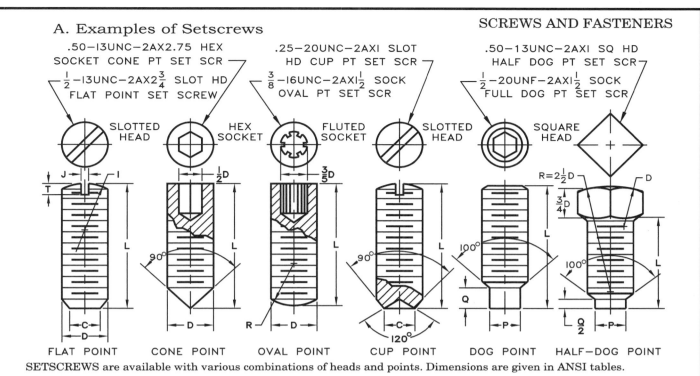

SETSCREWS are available with various combinations of heads and points. Dimensions are given in ANSI tables.

B. Standard Wood Screws

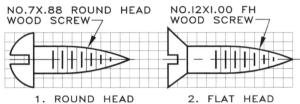

NO.7X.88 ROUND HEAD WOOD SCREW

NO.12X1.00 FH WOOD SCREW

1. ROUND HEAD

2. FLAT HEAD

Wood screws can be drawn approximately by referring to this background grid and dimensions from thread the tables.

C . Drilling and Tapping Holes

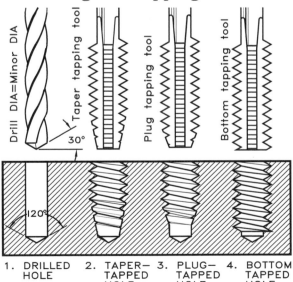

1. DRILLED HOLE   2. TAPER-TAPPED HOLE   3. PLUG-TAPPED HOLE   4. BOTTOM TAPPED HOLE

Drilling and threading interior holes can be performed in the four steps given above from drilling to bottom tapping (threading).

D. Examples of Bolts and Screws

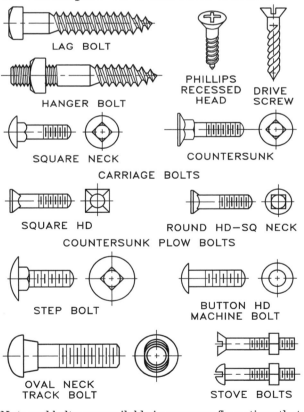

LAG BOLT

HANGER BOLT

PHILLIPS RECESSED HEAD

DRIVE SCREW

SQUARE NECK

COUNTERSUNK

CARRIAGE BOLTS

SQUARE HD

ROUND HD-SQ NECK

COUNTERSUNK PLOW BOLTS

STEP BOLT

BUTTON HD MACHINE BOLT

OVAL NECK TRACK BOLT

STOVE BOLTS

Nuts and bolts are available in many configurations that are not covered; only the more standard ones are introduced in this introductory coverage. Their usage is based the application at hand and the loads that they will be exposed to.

A. Standard Types of Keys and Keyways

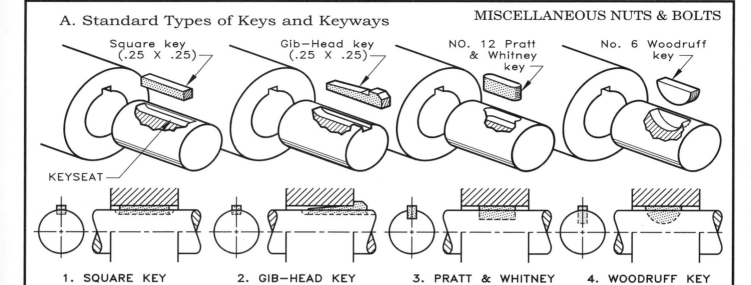

Square key (.25 X .25) Gib–Head key (.25 X .25) NO. 12 Pratt & Whitney key No. 6 Woodruff key

KEYSEAT

1. SQUARE KEY   **2. GIB–HEAD KEY**   **3. PRATT & WHITNEY**   **4. WOODRUFF KEY**

KEYS: The major types of keys and their keyways that are used to hold parts together are shown here.

B. Types of Lock Washers

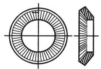

LOCK WASHER (Helical)

External   Internal
STAR WASHERS

COUNTERSUNK STAR WASHER

RIB WASHER (Section view)

Lock washers reduce the likelihood that threaded parts will loosen because of vibration and movement.

C. Examples of Pins

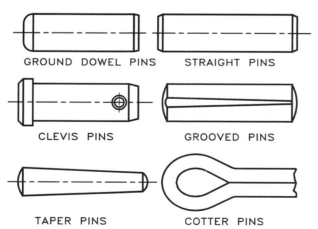

GROUND DOWEL PINS   STRAIGHT PINS

CLEVIS PINS   GROOVED PINS

TAPER PINS   COTTER PINS

Pins are used to hold parts together in assesmbly when it is desired that the parts can be unassembled.

D. Standard Grease Fittings

Thread size	$\frac{1}{8}$ 3 mm		$\frac{1}{4}$ 6 mm		$\frac{3}{8}$ 10 mm	
Overall length	L = in.	mm	L = in.	mm	L = in.	mm
Straight	.625	16	1.000	25	1.200	30
90° Elbow	.800	20	1.250	32	1.400	36
45° Angle	1.000	25	1.500	38	1.600	41

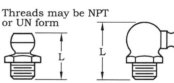

Threads may be NPT or UN form

1. STRAIGHT   **2. 90° ANGLE**   **3. 45° ANGLE**

Three standard types of grease fittings used to lubricate moving parts with a grease gun are shown here.

E. Double-Line Spring Construction

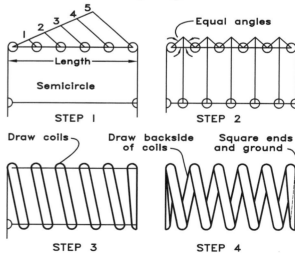

Equal angles

Length
Semicircle

STEP 1   **STEP 2**

Draw coils   Draw backside of coils   Square ends and ground

STEP 3   **STEP 4**

These steps illustrate the procedure in drawing a double-line representation of a spring. Springs can also be drawn with single lines but with less realism.

Graphics & Design
Miscellaneous Nuts & Bolts

FILE: NAME:
SEC: DATE: MIN:

GRADE

109B

APPENDIX 6
REGULAR HELICAL SPRING LOCK WASHERS (iNCHES)

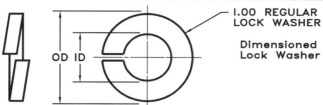

1.00 REGULAR LOCK WASHER

Dimensioned Lock Washer

SCREW SIZE	ID SIZE	OD SIZE	THICK-NESS	SCREW SIZE	ID SIZE	OD SIZE	THICK-NESS
0.164	0.168	0.175	0.040	0.812	0.826	1.367	0.203
0.190	0.194	0.202	0.047	0.875	0.890	1.464	0.219
0.216	0.221	0.229	0.056	0.938	0.954	1.560	0.234
0.250	0.255	0.263	0.062	1.000	1.017	1.661	0.250
0.312	0.318	0.328	0.078	1.062	1.080	1.756	0.266
0.375	0.382	0.393	0.094	1.125	1.144	1.853	0.281
0.438	0.446	0.459	0.109	1.188	1.208	1.950	0.297
0.500	0.509	0.523	0.125	1.250	1.271	2.045	0.312
0.562	0.572	0.587	0.141	1.312	1.334	2.141	0.328
0.625	0.636	0.653	0.156	1.375	1.398	2.239	0.344
0.688	0.700	0.718	0.172	1.438	1.462	2.334	0.359
0.750	0.763	0.783	0.188	1.500	1.525	2.430	0.375

METRIC LOCK WASHERS-DIN 127 (Millimeters)z

SCREW SIZE	ID SIZE	OD SIZE	THICK-NESS	SCREW SIZE	ID SIZE	OD SIZE	THICK-NESS
4	4.1	7.1	0.9	22	22.5	34.5	4
5	5.1	8.7	1.2	24	24.5	38.5	5
6	6.1	11.1	1.6	27	27.5	41.5	5
8	8.2	12.1	1.6	30	30.5	46.5	6
10	10.2	14.2	2	33	33.5	53.5	6
12	12.1	17.2	2.2	36	36.5	56.5	6
14	14.2	20.2	2.5	39	39.5	59.5	6
16	16.2	23.2	3	42	42.5	66.5	7
18	18.2	26.2	3.5	45	45.5	69.5	7
20	20.2	28.2	3.5	48	49	73	7

APPENDIX 7
SQUARE BOLTS (Inches)

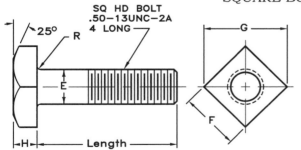

SQ HD BOLT .50-13UNC-2A 4 LONG

25°

STANDARD COMMERCIAL LENGTHS

*14 MEANS THAT LENGTHS ARE AVAILABLE AT 1 INCH INCREMENTS UP 14 INCHES.

DIA	E Max.	F Max.	G Avg.	H Max.	R Max.	DIA	E Max.	F Max.	G Avg.	H Max.	R Max.
1/4	.250	.375	.530	.188	.031	3/4	.750	1.125	1.591	.524	.062
5/16	.313	.500	.707	.220	.031	7/8	.875	1.313	1.856	.620	.062
3/8	.375	.563	.795	.268	.031	1	1.000	1.500	2.121	.684	.093
7/16	.438	.625	.884	.316	.031	1-1/8	1.125	1.688	2.386	.780	.093
1/2	.500	.750	1.061	.348	.031	1-1/4	1.250	1.875	2.652	.876	.093
5/8	.625	.938	1.326	.444	.062	1-3/8	1.375	2.625	2.917	.940	.093
						1-1/2	1.500	2.250	3.182	1.036	.093

APPENDIX 8
COTTER PINTS: AMERICAN NATIONAL STANDARD INSTITURE

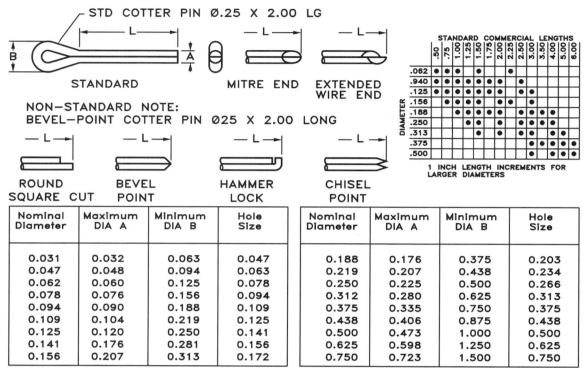

Nominal Diameter	Maximum DIA A	Minimum DIA B	Hole Size
0.031	0.032	0.063	0.047
0.047	0.048	0.094	0.063
0.062	0.060	0.125	0.078
0.078	0.076	0.156	0.094
0.094	0.090	0.188	0.109
0.109	0.104	0.219	0.125
0.125	0.120	0.250	0.141
0.141	0.176	0.281	0.156
0.156	0.207	0.313	0.172

Nominal Diameter	Maximum DIA A	Minimum DIA B	Hole Size
0.188	0.176	0.375	0.203
0.219	0.207	0.438	0.234
0.250	0.225	0.500	0.266
0.312	0.280	0.625	0.313
0.375	0.335	0.750	0.375
0.438	0.406	0.875	0.438
0.500	0.473	1.000	0.500
0.625	0.598	1.250	0.625
0.750	0.723	1.500	0.750

Courtesy of ANSI: B18.8.1−1983

APPENDIX 9
STRAIGHT PINS

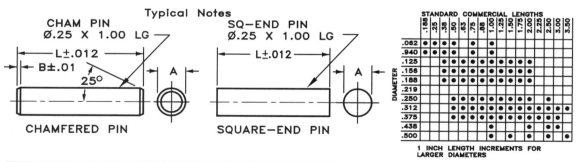

| Nominal DIA | Diameter A | | Chamfer B |
	Max	Min	
0.062	0.0625	0.0605	0.015
0.094	0.0937	0.0917	0.015
0.109	0.1094	0.1074	0.015
0.125	0.1250	0.1230	0.015
0.156	0.1562	0.1542	0.015
0.188	0.1875	0.1855	0.015

| Nominal DIA | Diameter A | | Chamfer B |
	Max	Min	
0.219	0.2187	0.2167	0.015
0.250	0.2500	0.2480	0.015
0.312	0.3125	0.3095	0.015
0.375	0.3750	0.3720	0.030
0.438	0.4345	0.4345	0.030
0.500	0.4970	0.4970	0.030

ANSI: B5.20

APPENDIX
110B

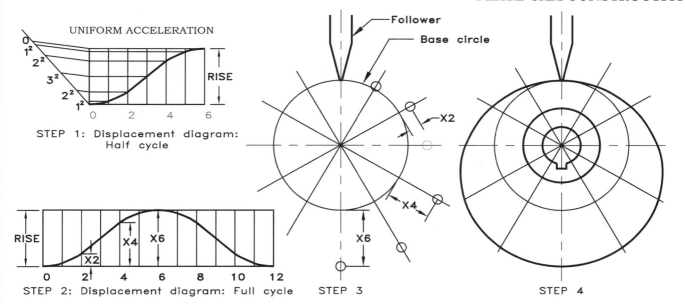

UNIFORM ACCELERATION

STEP 1: Displacement diagram: Half cycle

STEP 2: Displacement diagram: Full cycle

Follower

Base circle

X2

X4

X6

STEP 3

STEP 4

A PLATE CAM FOR UNIFORM ACCELERATION

STEP 1: Draw a displacement diagram with a vertical axis equal to the desired rise. Divide the X-axis into 12 equal divisions representing 30 degree intervals. Draw a construction line through 0, locate the 1^2, 2^2, 3^2 divisions, and project them to their respective vertical axes. Reverse this sequence (3^2, 2^2, 1^2) to locate the upper portion of the curve.

STEP 2: Continue the same construction to find the right half of the symmetrical curve by decreasing the values of the rise in reverse order.

STEP 3: Draw the base circle and knife-edge follower. Divide the circle into the same number of sectors as there are divisions in the displacement diagram. Transfer distances from the displacement diagram to their respective radial lines of the base circle and measure outward from the base circle.

STEP 4: Connect the points found in Step 3 with a smooth curve to complete the cam. Draw the cam hub and keyway.

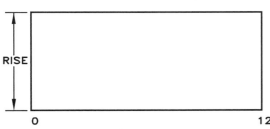

RISE

STEP 2: Displacement: Full Cycle

UNIFORM MOTION

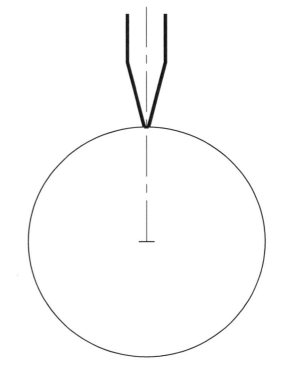

UNIFORM ACCELERATION CAM

STEP 1: Complete the displacement diagram (above) following the steps illustrated in the example above for a cam of this type with a maximum rise equal to the vertical distance of the diagram above.

STEP 2: By using the data in the displacement diagram, draw the plate cam, show your construction used in the process, label it as in the example, and show a hub for a 0.70 diameter shaft.

Graphics & Design

Assignment Solution

FILE: . . . . NAME: . . . .

SEC: . . . . DATE: . . . . MIN: . . . .

GRADE . . . .

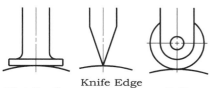

Flat Surface   Knife Edge   Roller

TYPES OF CAM FOLLOWERS

These are three types of cam followers that are used with plate cams.

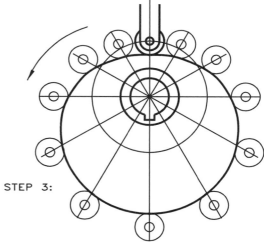

STEP 1: DISPLACEMENT DIAGRAM—HARMONIC

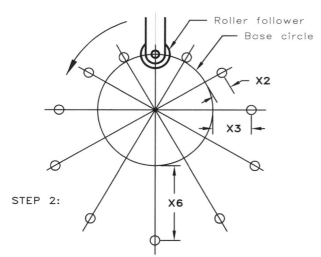

STEP 2:

A PLATE CAM FOR HARMONIC MOTION
STEP 1: Draw a semicircle on the vertical side of the displacement diagram whose diameter equals the rise of the follower. Divide the semicircle into the same number of segments as there are between 0 and 180 degrees on the horizontal axis of the diagram. Plot the displacement curve by measuring points (X1, X2 . . . X6) from the base circle.

STEP 3:

STEP 2: Draw the base circle and the circular follower, and divide the circle into the same number of divisions as on the displacement diagram. Transfer distances X1, X2, etc., from the displacement diagram to their respective radial lines of the circle measuring outward from it.

STEP 3: Draw circles to represent the positions of the roller as the cam revolves counterclockwise. Draw the cam profile tangent to all the roller positions of the cam.

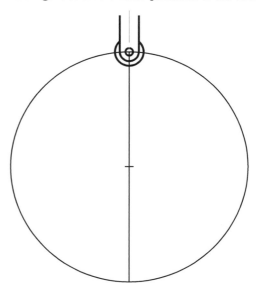

HARMONIC-MOTION CAM:
STEP 1: Complete the displacement diagram below following the steps illustrated in the example above for a cam of this type with a maximum rise equal to the vertical distance of the diagram below.

STEP 2: By using the data in the displacement diagram, draw the plate cam, show your construction used in this process, label it like in the example, and show the hub as well.

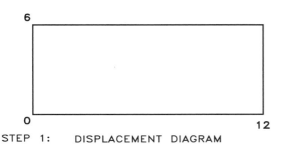

STEP 1:   DISPLACEMENT DIAGRAM

Graphics & Design
Plate Cam Construction

FILE: .... NAME: ....
SEC: .... DATE: .... MIN: ....

GRADE

112

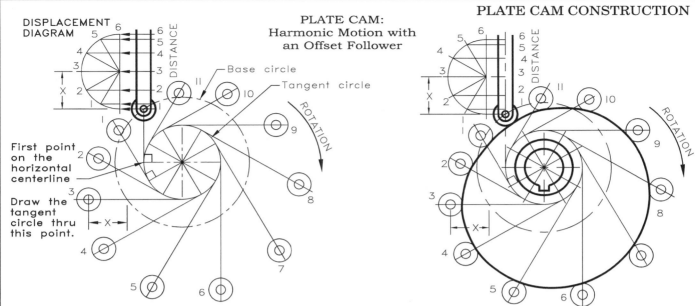

PLATE CAM:
Harmonic Motion with
an Offset Follower

A PLATE CAM FOR HARMONIC MOTION
STEP 1: The follower is an offset roller at the end of its
vertical arm that is limited to up and down motions. A semicircle, representing the harmonic rise of the follower, is drawn
upward from the follower's center. This semicircle is divided
into 30 degree increments, establishing the distances of the
follower's rise and fall at these intervals.

STEP 2: The vertical centerline of the follower is drawn downward to its intersection with the base circle's horizontal centerline to establish the diameter of the tangent circle. Other
locations of the follower are found by drawing lines tangent to
the tangent circle at 30° intervals, and drawing the rollers on
each location. Roller 3 is measured X-distance from the displacement diagram to the 3-position measured from the base circle.

STEP 3: Draw the cam to be tangent to the positions of the
rollers found in the previous construction. Additional
roller locations could have been found (at 15° intervals
for example) to increase the accuracy of drawing the cam's
outline.

STEP 4: Draw the cam's hub and its keyway for mounting
the cam on its shaft for proper use.

SCALE: 1=2"

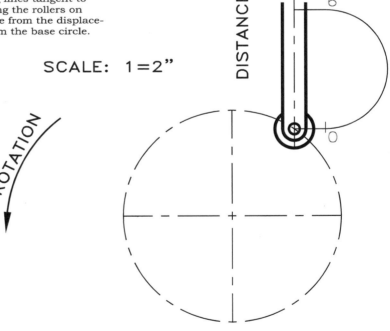

HARMONIC-MOTION CAM
STEP 1: By following the steps outlined
above, draw a cam with an offset follower
with the rise from 0 to 6 as indicated by
its semicircular displacement diagram.
Label your construction.

STEP 2: Draw a hub and keyway for a
shaft that is 1.30 inches in diameter.

FILE: .... NAME: ....

SEC: .... DATE: ....

MIN: ....

GRADE ....

Ø60

R5-2 PLACES

72

48

24

36REF   72   Ø25

126

22 LINK-1020 STEEL
2 REQ-HALF SIZE

SI

Letter Height=H=.125   Arrow=H   Place on edge of smoothed surface

32 REF
Optional reference dimension

3H

Ø20   R10
Diameter   Radius   Finish mark

Part name   Number required

24   BASE-2 REQ

4H
Part number

SI

SI symbol: Larger than letters

REQUIRED: Measure the features of the part with the metric scale asigned (1:10, 1:2, 1:20, etc.) and letter the dimensions, rounded to the nearest whole, even values. Give the two finish marks, the part number (24), part name (LEFT LINK), material (1020 STEEL), number required (2), and the scale assigned using the guidelines provided.

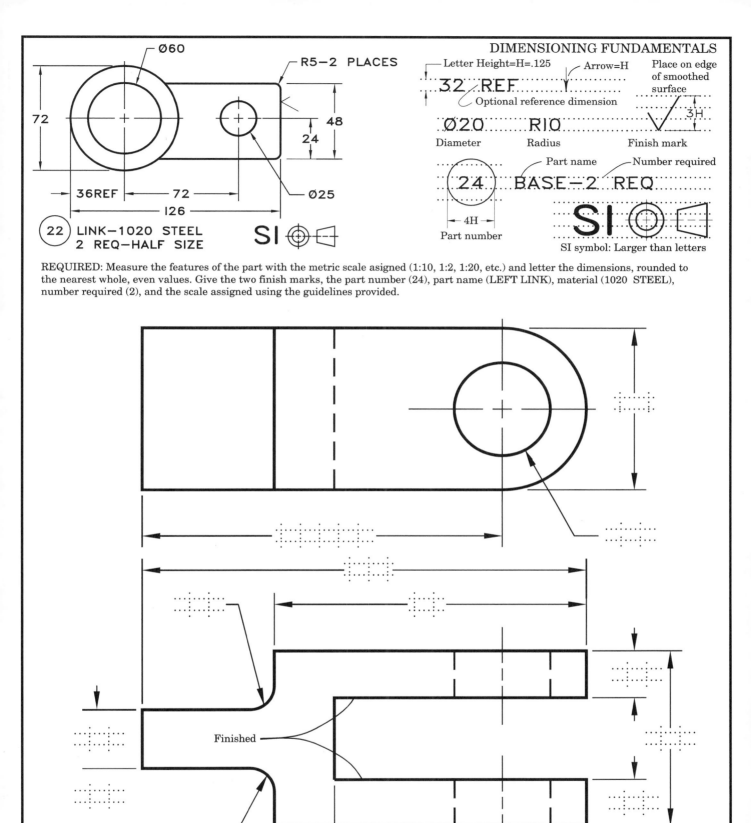

Finished

SI

SCALE:

Graphics & Design
Dimensioning Fundamentals

FILE: NAME:
SEC: DATE: MIN:

GRADE
114

Graphics & Design

Assignment Solution

FILE: NAME:

SEC: DATE:

MIN:

GRADE

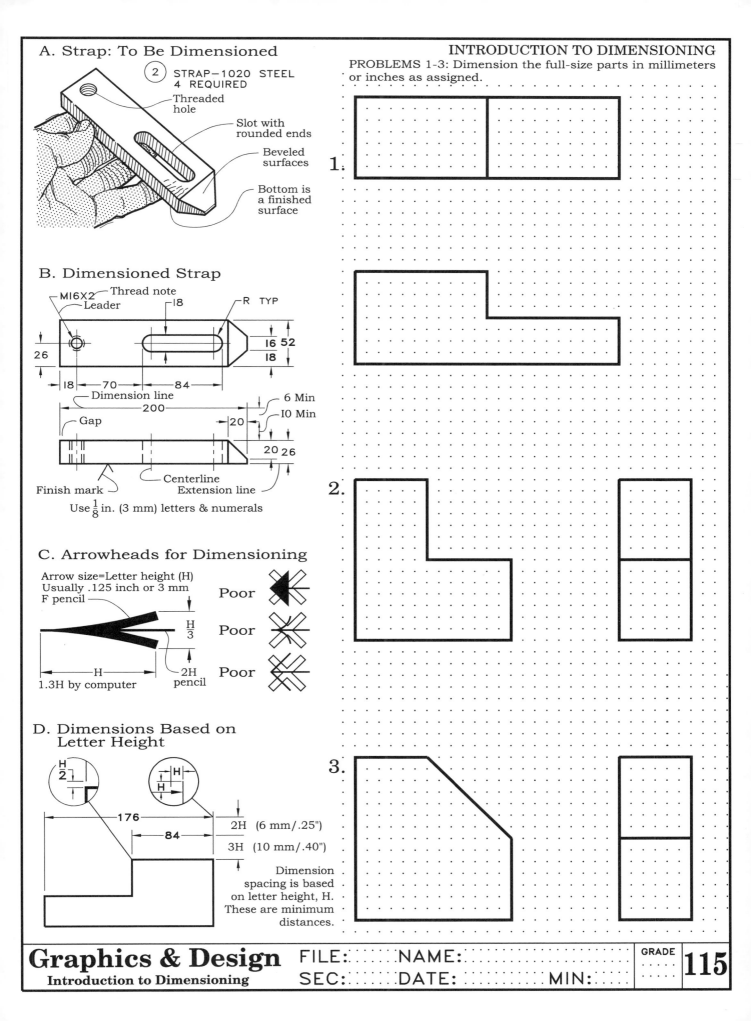

A. Strap: To Be Dimensioned

(2) STRAP—1020 STEEL
4 REQUIRED

- Threaded hole
- Slot with rounded ends
- Beveled surfaces
- Bottom is a finished surface

PROBLEMS 1-3: Dimension the full-size parts in millimeters or inches as assigned.

1:

B. Dimensioned Strap

- Thread note
- M16X2
- Leader
- 18
- R TYP
- 26
- 16 52
- 18
- 18 70 84
- Dimension line
- 200
- 6 Min
- 10 Min
- Gap
- 20
- 20 26
- Finish mark
- Centerline
- Extension line

Use 1/8 in. (3 mm) letters & numerals

2.

C. Arrowheads for Dimensioning

Arrow size=Letter height (H)
Usually .125 inch or 3 mm
F pencil

- H/3
- H
- 2H pencil

1.3H by computer

Poor

Poor

Poor

D. Dimensions Based on Letter Height

- H/2
- H
- H
- 176
- 84
- 2H (6 mm/.25")
- 3H (10 mm/.40")

Dimension spacing is based on letter height, H. These are minimum distances.

3.

6.4,6.4

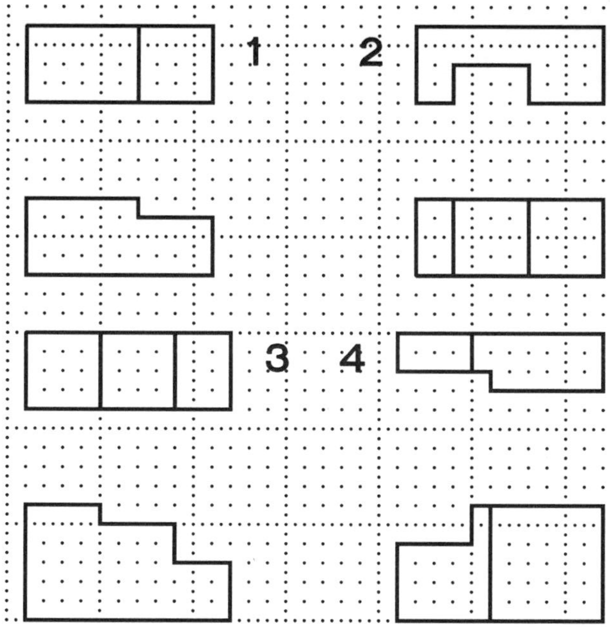

1

2

3

4

0,0

DIMENSIONING PRINCIPLES

PLOT AND DIMENSION THE FOUR PARTS.
USE DIMENSIONING TEXT THAT IS 0.10
INCHES HIGH. SCALE: FULL SIZE

Graphics & Design
Dimensioning Principles

FILE: .... NAME: ....

SEC: .... DATE: ....

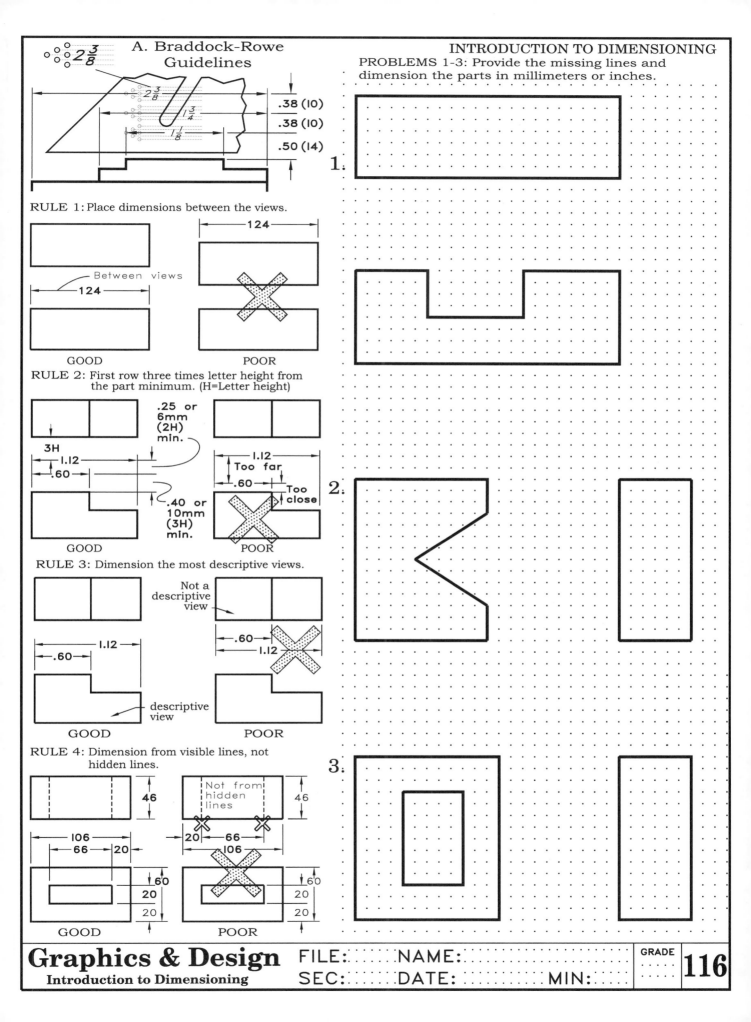

A. Braddock-Rowe
Guidelines

$2\frac{3}{8}$

$2\frac{3}{8}$

$1\frac{3}{4}$

$1\frac{1}{8}$

.38 (10)

.38 (10)

.50 (14)

PROBLEMS 1-3: Provide the missing lines and
dimension the parts in millimeters or inches.

1.

RULE 1: Place dimensions between the views.

124

Between views

124

GOOD

POOR

RULE 2: First row three times letter height from
the part minimum. (H=Letter height)

.25 or
6mm
(2H)
min.

3H

1.12

.60

.40 or
10mm
(3H)
min.

1.12
Too far
.60
Too
close

GOOD

POOR

RULE 3: Dimension the most descriptive views.

Not a
descriptive
view

1.12

.60

.60
1.12

descriptive
view

GOOD

POOR

RULE 4: Dimension from visible lines, not
hidden lines.

46

Not from
hidden
lines

46

106

66   20

60

20

20

20   66

106

60

20

20

GOOD

POOR

2.

3.

Graphics & Design
Introduction to Dimensioning

FILE: NAME:

SEC: DATE: MIN:

GRADE

116

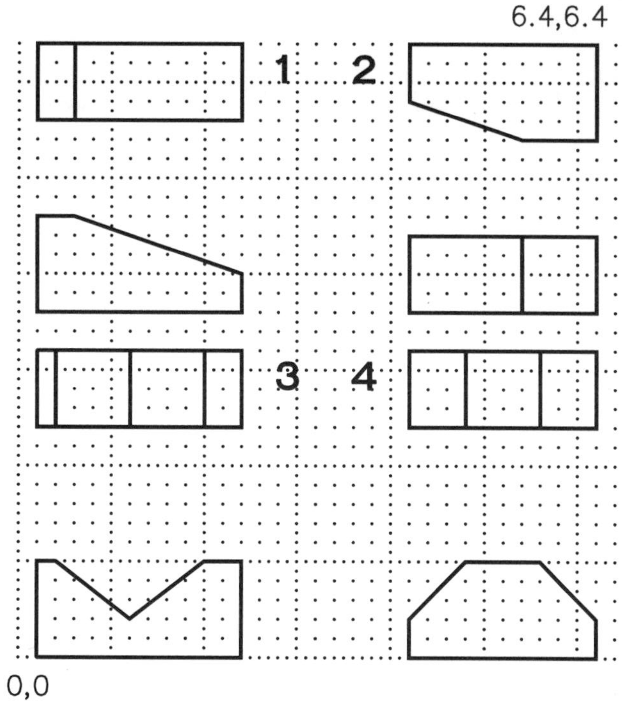

6.4,6.4

1   2

3   4

0,0

DIMENSIONING PRINCIPLES

PLOT AND DIMENSION THE PARTS. USE
DIMENSIONING TEXT THAT IS 0.08 IN.
HIGH. FULL SIZE

Graphics & Design
Dimensioning Principles

FILE: .... NAME: ....

SEC: .... DATE: ....

GRADE 116B

MIN: ....

RULE 5: Give an overall dimension and omit one of the chain dimensions.

GOOD — POOR — Omit one

RULE 6 (Deviation): If all chain dimensions are given, mark one of lesser importance as a reference dimension, REF.

Reference dimension marked "REF" — Reference dimension in parentheses

GOOD — GOOD

RULE 7: Organize and align dimensions for ease of reading.

Dimensions aligned and grouped. — Dimensions are not grouped or aligned.

GOOD — POOR

RULE 8: Do not repeat dimensions.

Errors may occur

GOOD — POOR

RULE 9: Dimension cylinders in their rectangular views with diameters

Not from circular view

Radius poor also.

GOOD — POOR

PROBLEMS 1-3: Provide the missing lines and dimension the parts in millimeters or inches.

1.

2.

3.

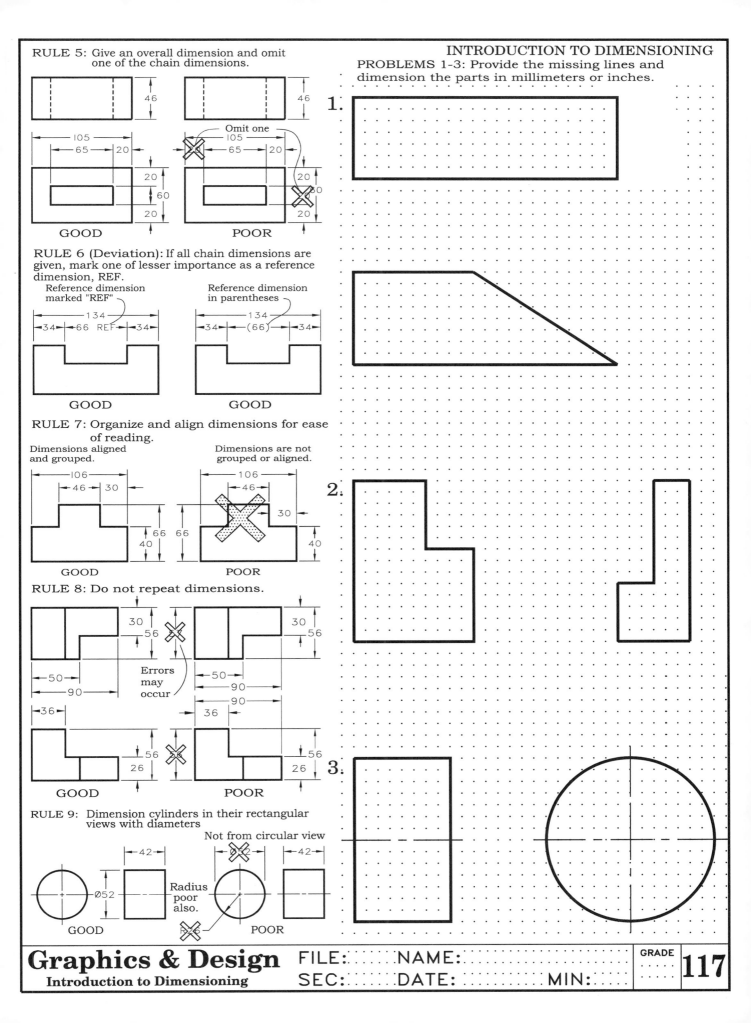

Graphics & Design
Introduction to Dimensioning

6.4,6.4

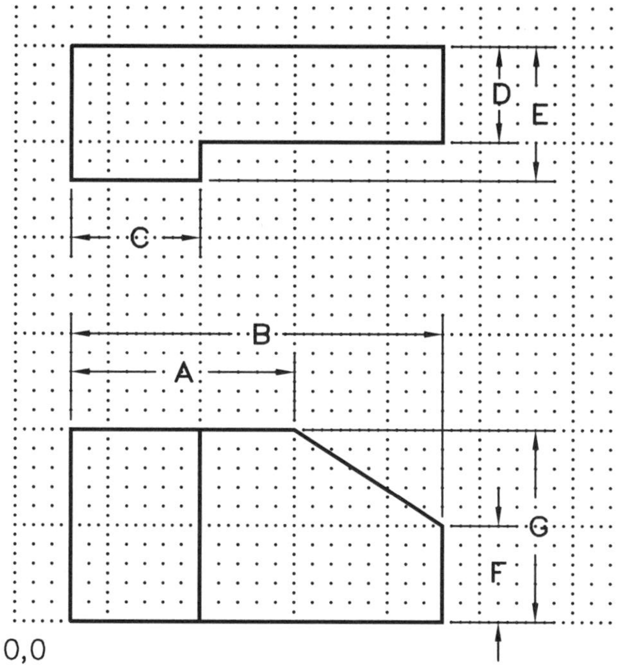

E
D

G
F

B
A
C

0,0

DIMENSIONING: METRIC

PLOT AND DIMENSION THE FULL-SIZE
PART IN MILLIMETERS WHERE EACH
SQUARE IS EQUAL TO 5 MM.
EXPERIMENT WITH VARIOUS SCALES.
FOR EXAMPLE, PLOT AND DIMENSION THE
PART TO BE DOUBLE SIZE, TRIPLE SIZE
AND SO FORTH.

Graphics & Design
Dimensioning: Metric

FILE: .... NAME: ....

SEC: .... DATE: ....

GRADE 117B

MIN: ....

RULE 10: Dimensions should not cross other lines.

1.12
.60

Dimension line crosses

.60
1.12

Start with shortest dimension

GOOD
POOR

RULE 11: Extension lines may cross other lines.

1.12
.60

Crossing extension lines OK

1.12
.60

Poor

.52
.34

.52
.34

GOOD
POOR

RULE 12: Do not place dimensions within the views unless absolutely necessary.

52
52

90
36 26

Poor

90
Close
26

20 70
30

70

50

Poor

GOOD
POOR

RULE 13: Continued Dimensions can be placed inside larger notches.

50
30

Better inside

40 20
80

30
20

40

80
60

Better outside

30
10

20 50

20

30

Inside of notch
Outside of notch

RULE 14: Dimension angles with coordinates or with a vertex and an angular arc.

70
50

20
Compass center

36°

GOOD
GOOD

PROBLEMS 1-3: Provide the missing lines and dimension the parts in millimeters or inches, as assigned.

1.

2.

3.

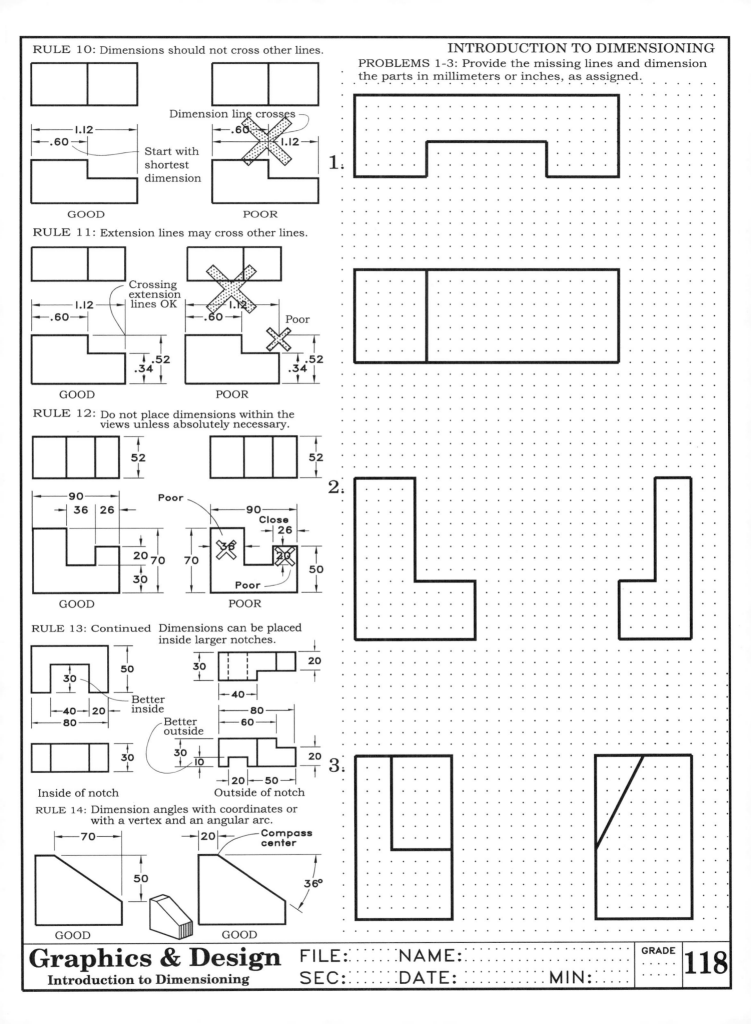

Graphics & Design
Introduction to Dimensioning

PROBLEMS 1-3: Provide the missing lines and dimension the parts in millimeters or inches as assigned.

RULE 15: Place angular dimensions outside the angle with extension lines.

Use extension lines to place arc outside the angle.

90°

GOOD

Dimension placed inside angle

90°

POOR

RULE 16: Dimension cylinders with diameters.

Dimension holes in circular views.

Ø26

Ø66

34

GOOD

Dimension cylinders in rectangular views with diameters.

Half view

20

6

Ø40

Ø66

GOOD

RULE 17: Hole sizes are best when given as diameters in circular views.

Holes noted

Ø12—4 HOLES

Ø52

GOOD

Hole diameter

OK when circular view is not given

Ø26—40 �

Ø26

OK, NOT BEST

RULE 18: Leaders should have horizontal elbows and point toward the hole's center.

Diameter sign

2.00 DIA

Ø52

GOOD

Do not use radius

R26

No elbow

Ø52 Stop at arc

POOR

RULE 19: Dimension arcs (less than 180°) with radii pointing to or from the center.

Arrow and dim. inside

R52

R52

Arrow inside dim. outside

R26

R26

Arrow and dim. outside

R12

R12

1.

2.

3.

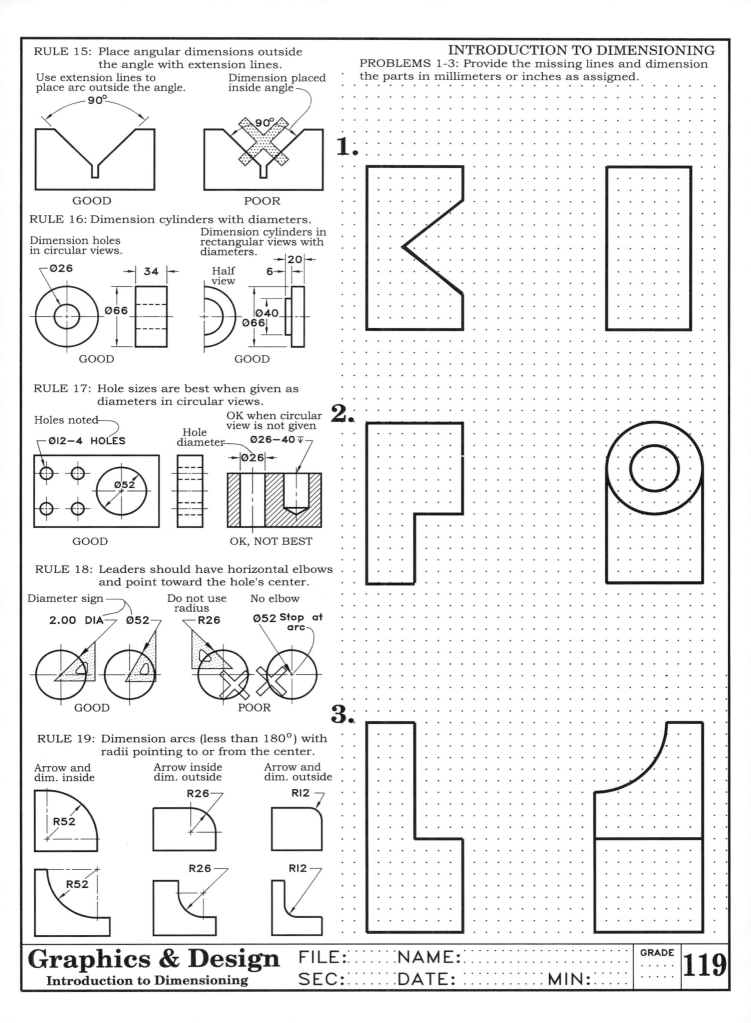

Graphics & Design
Introduction to Dimensioning

FILE: NAME:

SEC: DATE: MIN:

GRADE

119

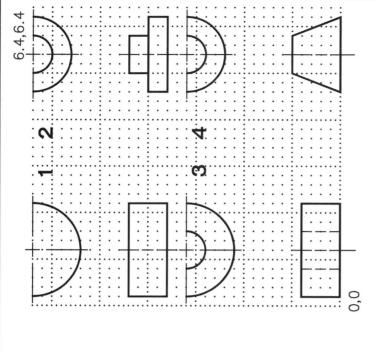

6.4,6.4

1 2

3 4

0,0

DIMENSIONING PRINCIPLES

PLOT AND DIMENSION THE PARTS. USE
DIMENSIONING TEXT THAT IS 0.08 IN.
HIGH. FULL SIZE

Graphics & Design
Dimensioning Principles

FILE:	NAME:
SEC:	DATE:

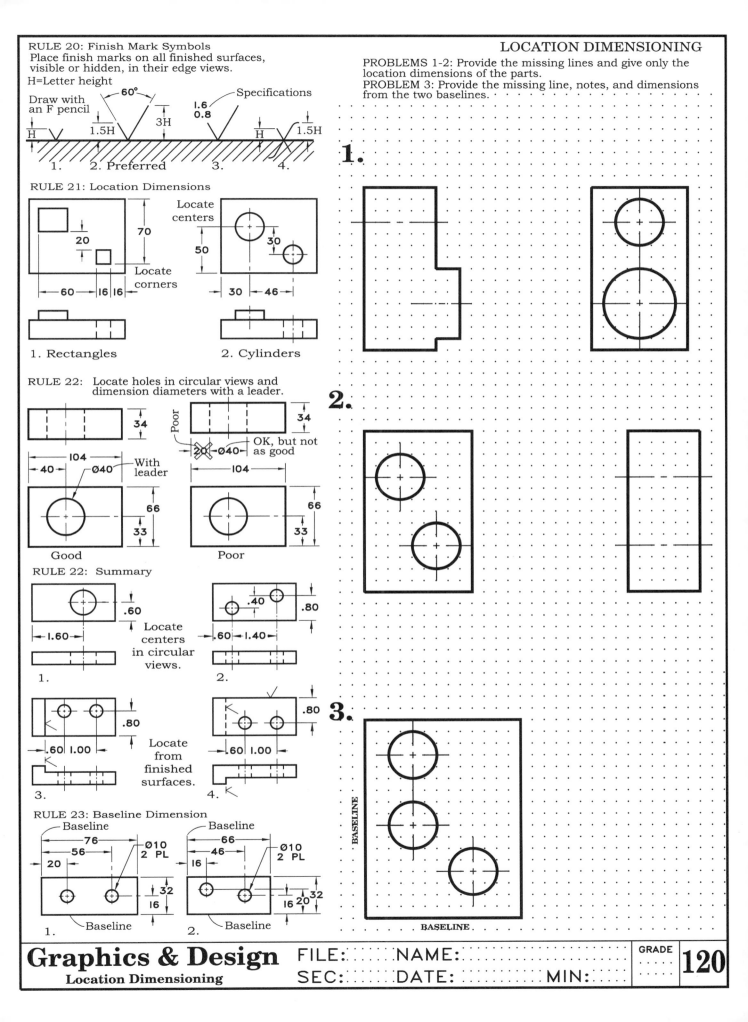

RULE 20: Finish Mark Symbols
Place finish marks on all finished surfaces,
visible or hidden, in their edge views.
H=Letter height

Draw with
an F pencil
60°
Specifications
1.6
0.8
3H
1.5H
H
1.5H
H
1. 2. Preferred 3. 4.

RULE 21: Location Dimensions

Locate
centers
70
20
50
30
Locate
corners
60 16 16
30 46

1. Rectangles 2. Cylinders

RULE 22: Locate holes in circular views and
dimension diameters with a leader.

34
Poor 34
OK, but not
20 Ø40 as good
104
40 Ø40
With
leader
104
66 66
33 33
Good Poor

RULE 22: Summary

.60
1.60
Locate
centers
in circular
views.
1.

.40 .80
.60 1.40
2.

.80
.60 1.00
Locate
from
finished
surfaces.
3.

.80
.60 1.00
4.

RULE 23: Baseline Dimension
Baseline Baseline
76 Ø10 66 Ø10
56 2 PL 46 2 PL
20 16
32 32
16 16 20
1. Baseline 2. Baseline

LOCATION DIMENSIONING

PROBLEMS 1-2: Provide the missing lines and give only the
location dimensions of the parts.
PROBLEM 3: Provide the missing line, notes, and dimensions
from the two baselines.

1.

2.

3.

BASELINE

BASELINE

Graphics & Design
Location Dimensioning

FILE: NAME:
SEC: DATE: MIN:

GRADE **120**

6.4,6.4

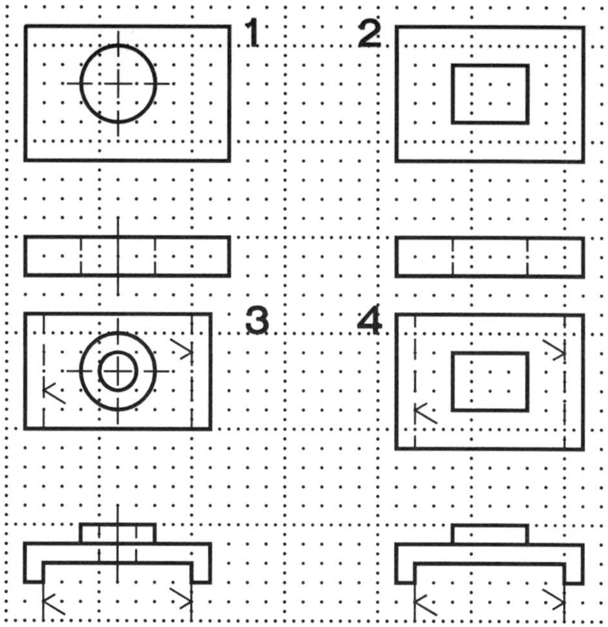

1　2

3　4

0,0

LOCATION DIMENSIONS

PLOT THE VIEWS AND GIVE THE
LOCATION DIMENSIONS ONLY. USE
TEXT THAT IS 0.08 IN. HIGH.

Graphics & Design
Location Dimensions

FILE: NAME:
SEC: DATE:

GRADE 120B

MIN:

A. Location of Holes

D. Circular Features

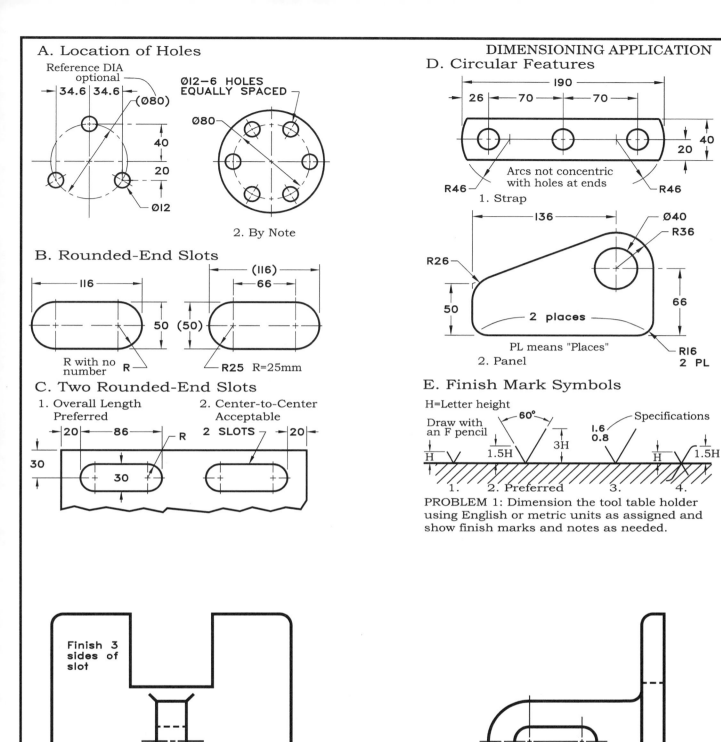

Reference DIA optional

34.6 | 34.6

(Ø80)

Ø12

40

20

Ø12–6 HOLES EQUALLY SPACED

Ø80

2. By Note

190

26 | 70 | 70

40

20

Arcs not concentric with holes at ends

R46

R46

1. Strap

B. Rounded-End Slots

116

50

R with no number

R

(116)

66

(50)

R25 R=25mm

136

Ø40

R36

R26

50

2 places

66

R16
2 PL

PL means "Places"

2. Panel

C. Two Rounded-End Slots

1. Overall Length Preferred

2. Center-to-Center Acceptable

20 | 86

R

30

30

2 SLOTS

20

E. Finish Mark Symbols

H=Letter height

Draw with an F pencil

60°

Specifications

1.6
0.8

H

1.5H

3H

H

1.5H

1.   2. Preferred   3.   4.

PROBLEM 1: Dimension the tool table holder using English or metric units as assigned and show finish marks and notes as needed.

Finish 3 sides of slot

Finish this surface

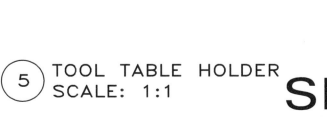

5   TOOL TABLE HOLDER
SCALE: 1:1

SI

Graphics & Design
Dimensioning Application

FILE: . . . . NAME: . . . .

SEC: . . . . DATE: . . . . MIN: . . . . . .

GRADE

121

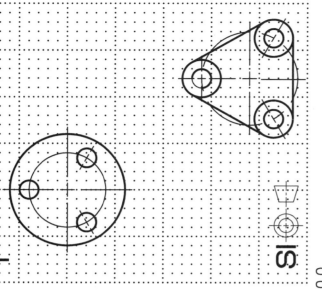

6,4,6,4

1

2

SI

0,0

DIMENSIONING HOLES

DIMENSION THE PARTS IN MILLIMETERS.
EACH SQUARE IS EQUAL TO 6 mm.
BOTH PLATES ARE 8 mm THICK.

Graphics & Design
Dimensioning Holes

FILE: .... NAME: ....

SEC: .... DATE: ....

A. Counterdrilled (CDRILL) Hole

Ø20 THRU — or CDRILL
Ø30 COUNTERDRILL
18 DEEP

Drill bit

30°   30°
120° Point

B. Countersunk (CSK) Hole

Ø.50–82° CSK
Ø1.00–3 HOLES
EQ SPACED

82° counter-sinking drill.

For screw heads

1. Multiple Holes   2. Countersink Tool

C. Spotfaced (SF) Hole

Ø20 THRU — or SF
Ø38 SPOTFACE — By words

By symbols

Ø16 THRU
⌴ Ø34 ⍌ 3

Spotfacing tool
Purpose is to smooth the surface. Depth is left to shop.

1. Section   2. View

D. Counterbored (CBORE) Holes

Ø16 THRU– Ø26 CBORE
10 DEEP–2 HOLES
By words

Ø20 THRU
⌴ Ø32 ⍌10
By symbols

Cbore tool

1. Counterbore: Section   2. Cbore: View

DIMENSIONING MACHINED HOLES

PROBLEM 1: Attach a note to the part below for a 30 mm DIA thru hole and a 45 mm DIA counterdrill 18 mm deep drawn as both a view and a full section through the hole.

1.

VIEW   FULL SECTION

PROBLEM 2: Same as problem 1 with these specifications: Drill thru with a 20 mm DIA drill, countersink with an 82° CSK to a 30 mm diameter.

2.

VIEW   FULL SECTION

PROBLEM 3: Same as problem 1 with these specifications: Drill thru with a 30 mm DIA drill, spotface with 60 mm DIA to about a depth of 3 mm.

PROBLEM 4: Same as problem 1 with these specifications: Drill thru with a 20 mm DIA drill, counterbore with 32 mm DIA to a depth of 10 mm.

3.   4.

FULL SECTION   FULL SECTION

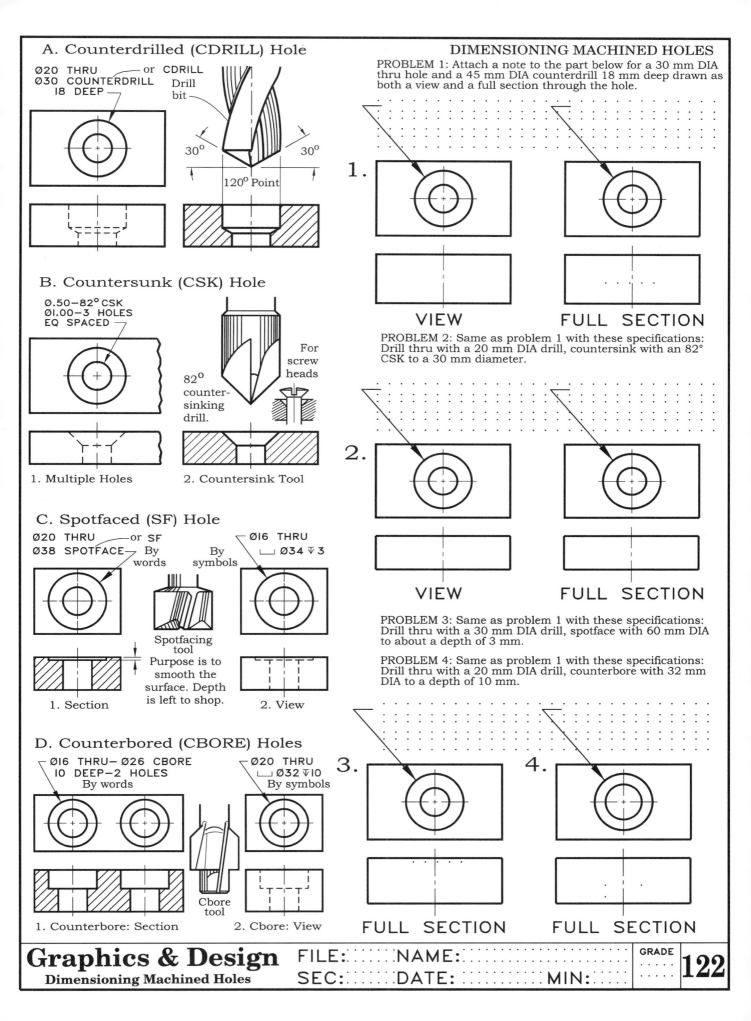

Graphics & Design
Dimensioning Machined Holes

FILE: .... NAME:

SEC: .... DATE: .... MIN:

GRADE

122

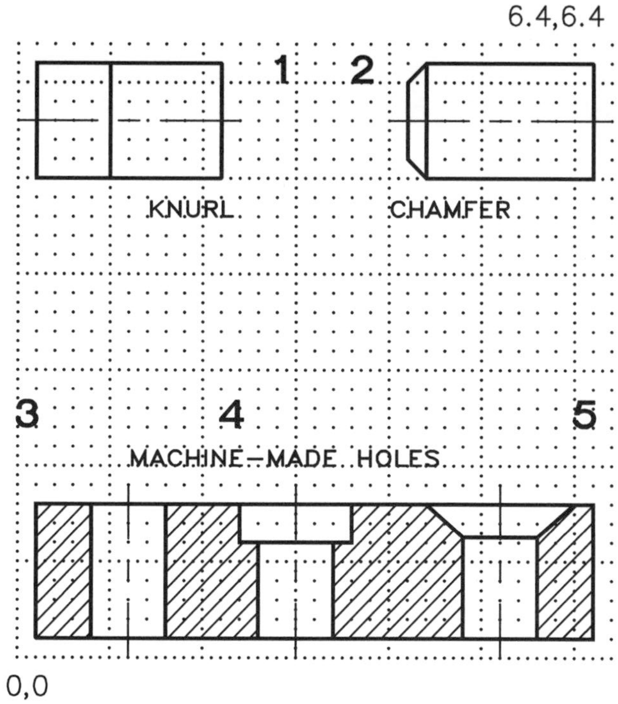

6.4,6.4

1   2

KNURL   CHAMFER

3   4   5

MACHINE-MADE HOLES

0,0

DIMENSIONING NOTES

PROVIDE MACHINING NOTES ON THE
PARTS ABOVE IN ACCORDANCE WITH
THE SPECIFICATIONS BELOW:

1. DRAW KNURLING SYMBOLS AT THE
   RIGHT END OF THE SHAFT AND
   GIVE A NOTE FOR A 64 DP DIAMOND
   KNURL.

2. DIMENSION THE CHAMFER.

3. GIVE NOTES THAT WILL DIMENSION
   THE MACHINE-MADE HOLES.

Graphics & Design
Dimensioning Notes

FILE: .... NAME: ....

SEC: .... DATE: ....

A. Machining Holes

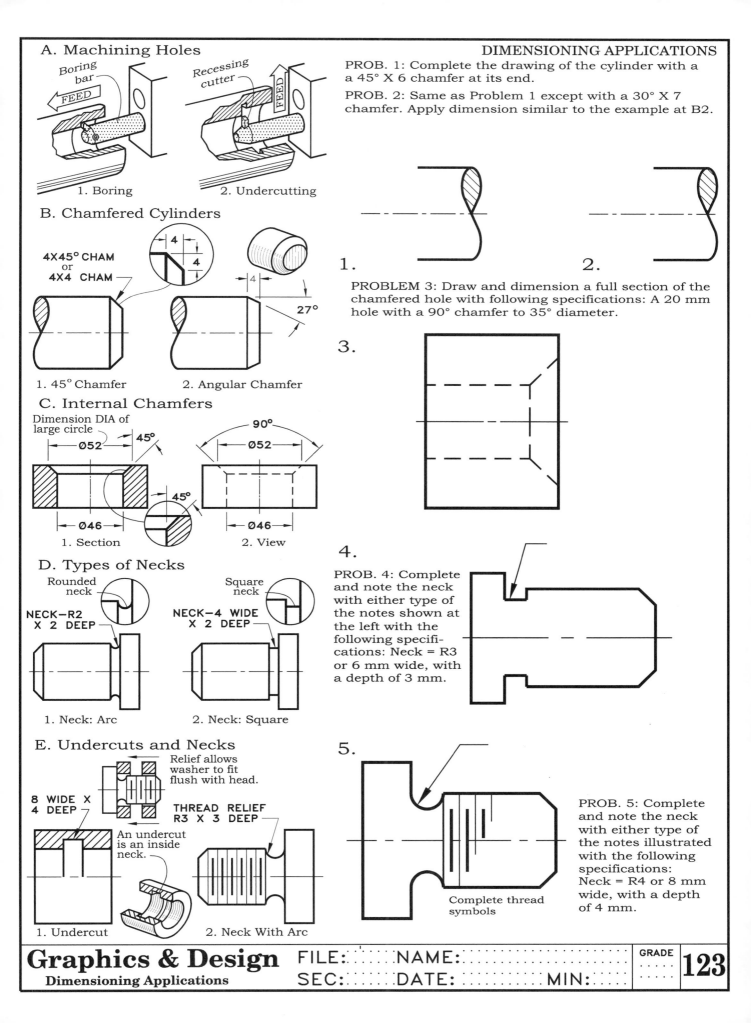

Boring bar
FEED

1. Boring

Recessing cutter
FEED

2. Undercutting

B. Chamfered Cylinders

4X45° CHAM
or
4X4 CHAM

4
4

4

27°

1. 45° Chamfer

2. Angular Chamfer

C. Internal Chamfers

Dimension DIA of large circle

Ø52   45°

Ø46

45°

1. Section

90°

Ø52

Ø46

2. View

D. Types of Necks

Rounded neck

NECK—R2 X 2 DEEP

1. Neck: Arc

Square neck

NECK—4 WIDE X 2 DEEP

2. Neck: Square

E. Undercuts and Necks

Relief allows washer to fit flush with head.

8 WIDE X 4 DEEP

An undercut is an inside neck.

THREAD RELIEF R3 X 3 DEEP

1. Undercut

2. Neck With Arc

PROB. 1: Complete the drawing of the cylinder with a a 45° X 6 chamfer at its end.

PROB. 2: Same as Problem 1 except with a 30° X 7 chamfer. Apply dimension similar to the example at B2.

1.

2.

PROBLEM 3: Draw and dimension a full section of the chamfered hole with following specifications: A 20 mm hole with a 90° chamfer to 35° diameter.

3.

4.

PROB. 4: Complete and note the neck with either type of the notes shown at the left with the following specifications: Neck = R3 or 6 mm wide, with a depth of 3 mm.

5.

PROB. 5: Complete and note the neck with either type of the notes illustrated with the following specifications: Neck = R4 or 8 mm wide, with a depth of 4 mm.

Complete thread symbols

Graphics & Design
Dimensioning Applications

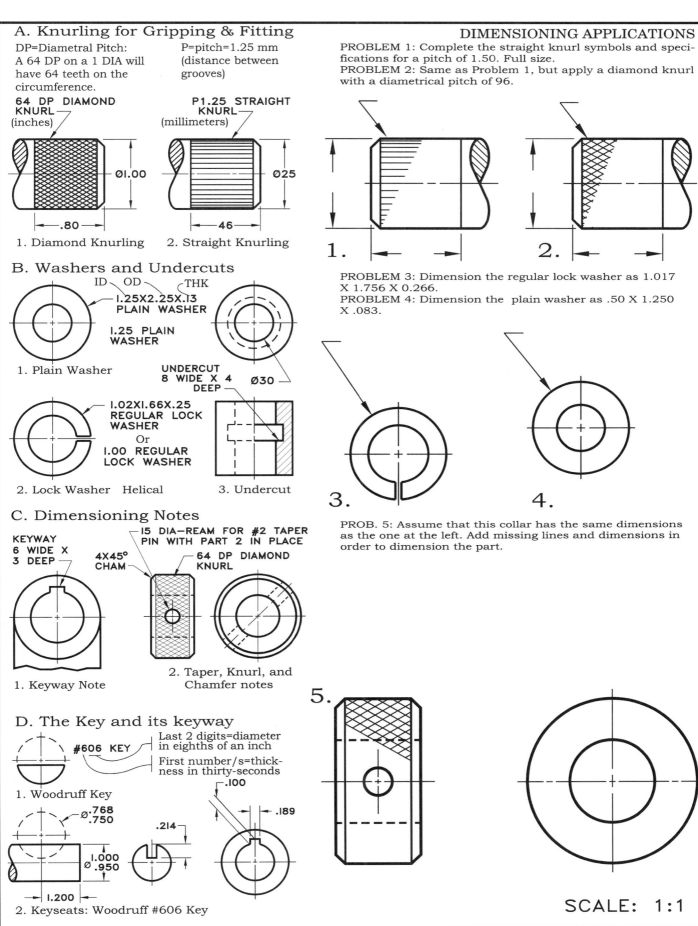

A. Knurling for Gripping & Fitting

DP=Diametral Pitch:
A 64 DP on a 1 DIA will
have 64 teeth on the
circumference.

P=pitch=1.25 mm
(distance between
grooves)

64 DP DIAMOND KNURL (inches)

P1.25 STRAIGHT KNURL (millimeters)

Ø1.00

Ø25

.80

46

1. Diamond Knurling
2. Straight Knurling

B. Washers and Undercuts

ID   OD   THK

1.25X2.25X.13
PLAIN WASHER

1.25 PLAIN
WASHER

1. Plain Washer

UNDERCUT
8 WIDE X 4
DEEP

Ø30

1.02X1.66X.25
REGULAR LOCK
WASHER
Or
1.00 REGULAR
LOCK WASHER

2. Lock Washer Helical

3. Undercut

C. Dimensioning Notes

KEYWAY
6 WIDE X
3 DEEP

15 DIA—REAM FOR #2 TAPER
PIN WITH PART 2 IN PLACE

4X45°
CHAM

64 DP DIAMOND
KNURL

1. Keyway Note

2. Taper, Knurl, and
Chamfer notes

D. The Key and its keyway

#606 KEY

Last 2 digits=diameter
in eighths of an inch

First number/s=thick-
ness in thirty-seconds

1. Woodruff Key

Ø.768
.750

.214

.100

.189

1.000
.950

1.200

2. Keyseats: Woodruff #606 Key

DIMENSIONING APPLICATIONS

PROBLEM 1: Complete the straight knurl symbols and specifications for a pitch of 1.50. Full size.
PROBLEM 2: Same as Problem 1, but apply a diamond knurl
with a diametrical pitch of 96.

1.

2.

PROBLEM 3: Dimension the regular lock washer as 1.017
X 1.756 X 0.266.
PROBLEM 4: Dimension the plain washer as .50 X 1.250
X .083.

3.

4.

PROB. 5: Assume that this collar has the same dimensions
as the one at the left. Add missing lines and dimensions in
order to dimension the part.

5.

SCALE: 1:1

Graphics & Design
Dimensioning Applications

FILE: NAME:
SEC: DATE: MIN:

GRADE

124

APPENDIX 10:
WOODRUFF KEYS AND KEYWAYS

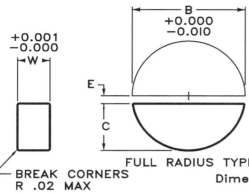

FULL RADIUS TYPE   FLAT BOTTOM TYPE

Dimensions are in inches

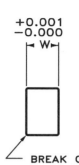

BREAK CORNERS
R .02 MAX

BREAK CORNERS
R .02 MAX

Key No.	W X B	C Max.	D Max.	E
204	1/16 X 1/2	.203	.194	.047
304	3/32 X 1/2	.203	.194	.047
404	1/8 X 1/2	.203	.194	.047
305	3/32 X 5/8	.250	.240	.063
405	1/8 X 5/8	.250	.240	.063
505	5/32 X 5/8	.250	.240	.063
406	1/8 X 3/4	.313	.303	.063

Key No.	W X B	C Max.	D Max.	E
506	5/32 X 3/4	.313	.303	.063
606	3/16 X 3/4	.313	.303	.063
507	5/32 X 7/8	.375	.365	.063
607	3/16 X 7/8	.375	.365	.063
807	1/4 X 7/8	.375	.365	.063
608	3/16 X 1	.438	.428	.063
609	3/16 X 1-1/8	.484	.475	.078

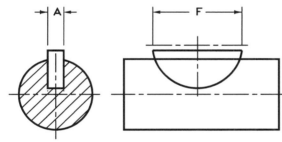

KEYSEAT—SHAFT

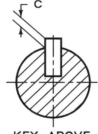

KEY ABOVE
SHAFT

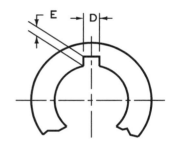

KEYSEAT—HUB

Key No.	A Min.	C +.005 −.000	F	D +.005 −.000	E +.005 −.000
204	.0615	.0312	.500	.0635	.0372
304	.0928	.0469	.500	.0948	.0529
404	.1240	.0625	.500	.1260	.0685
305	.0928	.0625	.625	.0948	.0529
405	.1240	.0469	.625	.1260	.0685
505	.1553	.0625	.625	.1573	.0841
406	.1240	.0781	.750	.1260	.0685

Key No.	A Min.	C +.005 −.000	F	D +.005 −.000	E +.005 −.000
506	.1553	.0781	.750	.1573	.0841
606	.1863	.0937	.750	.1885	.0997
507	.1553	.0781	.875	.1573	.0841
607	.1863	.0937	.875	.1885	.0997
807	.2487	.1250	.875	.2510	.1310
608	.1863	.3393	1.000	.1885	.0997
609	.1863	.3853	1.125	.1885	.0997

KEY SIZES VS. SHAFT SIZES

Shaft DIA	to .375	to .500	to .750	to 1.313	to 1.188	to 1.448	to 1.750	to 2.125	to 2.500
Key Nos.	204	304 305	404 405 406	505 506 507	606 607 608 609	807 808 809	810 811 812	1011 1012	1211 1212

APPENDIX 11: STANDARD KEYS AND KEYWAYS

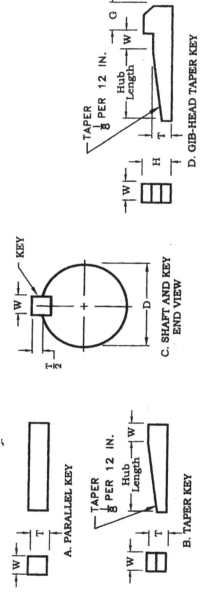

A. PARALLEL KEY

TAPER $\frac{1}{8}$ PER 12 IN. — Hub Length — B. TAPER KEY

KEY — C. SHAFT AND KEY END VIEW

TAPER $\frac{1}{8}$ PER 12 IN. — Hub Length — D. GIB-HEAD TAPER KEY

Sprocket Bore (= Shaft Diam.) Inches D	Keyway For Square Key Width W × Depth T/2	Keyway For Flat Key Width W × Depth T/2	Key Square Width W × Height T	Key Flat Width W × Height T	Tolerance on W and T (−)	Gib Head Square Key H	Gib Head Square Key G	Gib Head Flat Key H	Gib Head Flat Key G	Key Tol W (−)	Key Tol T (−)
1/2–9/16	1/8 × 1/16	1/8 × 3/64	1/8 × 1/8	1/8 × 3/32	0.002	1/4	7/32	3/16	1/8	0.002	0.002
5/8–7/8	3/16 × 3/32	3/16 × 1/16	3/16 × 3/16	3/16 × 1/8	0.002	5/16	9/32	1/4	3/16	0.002	0.002
13/16–1 1/4	1/4 × 1/8	1/4 × 3/32	1/4 × 1/4	1/4 × 3/16	0.002	7/16	11/32	5/16	1/4	0.002	0.002
1 5/16–1 3/8	5/16 × 5/32	5/16 × 1/8	5/16 × 5/16	5/16 × 1/4	0.002	9/16	13/32	3/8	5/16	0.002	0.002
1 7/16–1 3/4	3/8 × 3/16	3/8 × 3/16	3/8 × 3/8	3/8 × 1/4	0.002	11/16	15/32	7/16	3/8	0.002	0.002
1 13/16–2 1/4	1/2 × 1/4	1/2 × 3/16	1/2 × 1/2	1/2 × 3/8	0.0025	7/8	19/32	5/8	1/2	0.0025	0.0025
2 5/16–2 3/4	5/8 × 5/16	5/8 × 7/32	5/8 × 5/8	5/8 × 7/16	0.0025	1 1/16	23/32	3/4	5/8	0.0025	0.0025
2 7/8–3 1/4	3/4 × 3/8	3/4 × 1/4	3/4 × 3/4	3/4 × 1/2	0.0025	1 1/4	7/8	7/8	3/4	0.0025	0.0025
3 3/8–3 3/4	7/8 × 7/16	7/8 × 5/16	7/8 × 7/8	7/8 × 5/8	0.003	1 1/2	1	1 1/16	7/8	0.003	0.003
3 7/8–4 1/2	1 × 1/2	1 × 3/8	1 × 1	1 × 3/4	0.003	1 3/4	1 3/16	1 1/4	1	0.003	0.003
4 3/4–5 1/2	1 1/4 × 5/8	1 1/4 × 7/16	1 1/4 × 1 1/4	1 1/4 × 7/8	0.003	2	1 7/16	1 1/2	1 1/4	0.003	0.003
5 3/4–7 7/8	1 1/2 × 3/4	1 1/2 × 1/2	1 1/2 × 1 1/2	1 1/2 × 1	0.003	2 1/2	1 3/4	1 3/4	1 1/2	0.003	0.003
7 1/2–9 7/8	1 3/4 × 7/8	- - -	1 3/4 × 1 3/4	- - -	0.004	3	2	1 3/4	1 1/2	0.004	0.004
10–12 1/2	2 × 1	- - -	2 × 2	- - -	0.004	3 1/2	2 3/8	1 3/4	1 1/2	0.004	0.004

Standard Keyway Tolerances:

Straight Keyway—Width (W) + .005 / −.000   Depth (T/2) + .010 / −.000

Taper Keyway—Width (W) + .005 / −.000   Depth (T/2) + .000 / −.010

APPENDIX 12:
PLAIN AND WIDE WASHERS (Inches)

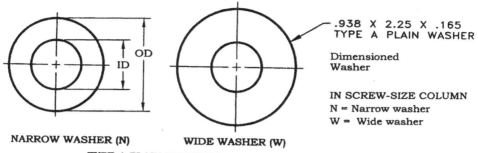

.938 X 2.25 X .165
TYPE A PLAIN WASHER

Dimensioned
Washer

IN SCREW-SIZE COLUMN
N = Narrow washer
W = Wide washer

NARROW WASHER (N)   WIDE WASHER (W)

TYPE A PLAIN WASHERS

Screw Size	ID Size	OD Size	Thick-ness	Screw Size	ID Size	OD Size	Thick-ness
0.138	0.156	0.375	0.049	0.875 N	0.938	1.750	0.134
0.164	0.188	0.438	0.049	0.875 W	0.938	2.250	0.165
0.190	0.219	0.500	0.049	1.000 N	1.062	2.000	0.134
0.188	0.250	0.562	0.049	1.000 W	1.062	2.500	0.165
0.216	0.250	0.562	0.065	1.125 N	1.250	2.250	0.134
0.250 N	0.281	0.625	0.065	1.125 W	1.250	2.750	0.165
0.250 W	0.312	0.734	0.065	1.250 N	1.375	2.500	0.165
0.312 N	0.344	0.688	0.065	1.250 W	1.375	3.000	0.165
0.312 W	0.375	0.875	0.083	1.375 N	1.500	2.750	0.165
0.375 N	0.406	0.812	0.065	1.375 W	1.500	3.250	0.180
0.375 W	0.438	1.000	0.083	1.500 N	1.625	3.000	0.165
0.438 N	0.469	0.922	0.065	1.500 W	1.625	3.500	0.180
0.438 W	0.500	1.250	0.083	1.625	1.750	3.750	0.180
0.500 N	0.531	1.062	0.095	1.750	1.875	4.000	0.180
0.500 W	0.562	1.375	0.109	1.875	2.000	4.250	0.180
0.562 N	0.594	1.156	0.095	2.000	2.125	4.500	0.180
0.562 W	0.594	1.469	0.190	2.250	2.375	4.750	0.220
0.625 N	0.625	1.312	0.095	2.500	2.625	5.000	0.238
0.625 N	0.625	1.750	0.134	2.750	2.875	5.250	0.259
0.750 W	0.812	1.469	0.134	3.000	3.125	5.500	0.284
0.750 W	0.812	2.000	0.148				

Dimensions are in inches.

APPENDIX 13:
METRIC FLAT AND WROUGHT WASHERS (Millimeters)

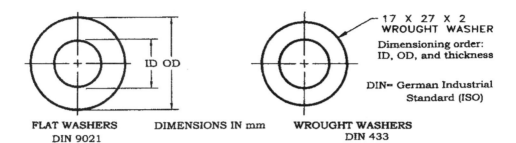

FLAT WASHERS
DIN 9021

DIMENSIONS IN mm

WROUGHT WASHERS
DIN 433

17 X 27 X 2
WROUGHT WASHER

Dimensioning order:
ID, OD, and thickness

DIN= German Industrial
Standard (ISO)

Screw Size	ID Size	OD Size	Thick-ness	
3	3.2	9	0.8	
4	4.3	12	1	
5	5.3	15	1.5	
6	6.4	18	1.5	
8	8.4	25	2	FLAT
10	10.5	30	2.5	WASHERS
12	13	40	3	
14	15	45	3	
16	17	50	3	
18	19	56	4	
20	21	60	4	
2.6	2.8	5.5	0.5	
3	3.2	6	0.5	
4	4.3	8	0.5	
5	5.3	10	1.0	
6	6.4	11	1.5	
8	8.4	15	1.5	WROUGHT
10	10.5	18	1.5	WASHERS
12	13	20	2.0	
14	15	25	2.0	
16	17	27	2.0	
18	19	30	2.5	
20	21	33	2.5	

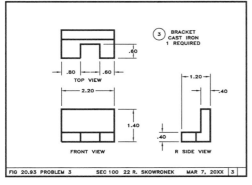

INSTRUCTIONS:
Draw and dimension the assigned problems on A-size sheets at a scale where the grid is .20 inches (or 5 millimeters). You will need to vary the spacing between the views to provide room for the dimensions. It is recommended that you sketch the views on scratch paper in advance to determine the required space needed for your final solution. Dimension in either inches or millimeters as assigned. Lines and centerlines may be missing in all views.

The example at the left is an instrument-drawn working drawing which is more accurate than freehand sketched versions.

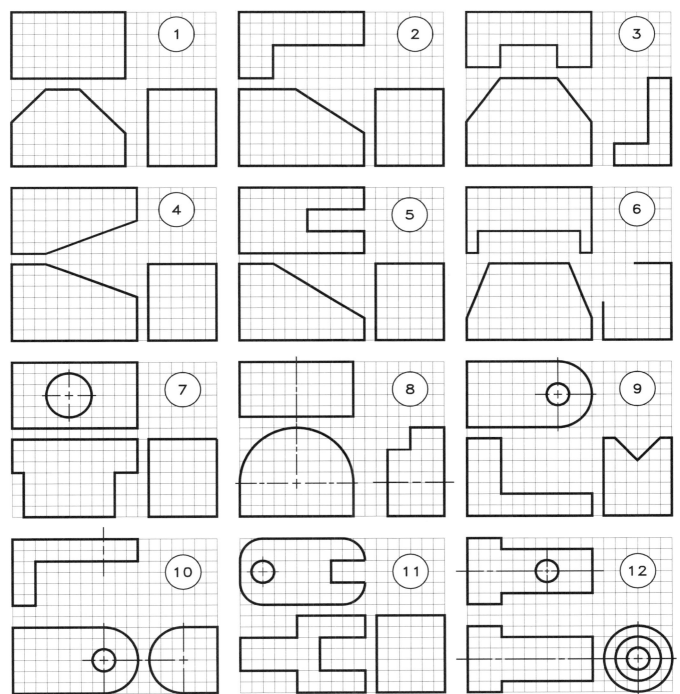

Graphics & Design
Dimensioning Applications

FILE: NAME:
SEC: DATE: MIN:

GRADE

127

TOLERANCES
±.032

R.05

.80 CBORE
.40 DEEP

Ø.600 Ø.60 250
Ø.598

Ø1.80

5

1.30
1.90

GLAND
1020 STEEL

FREEHAND WORKING DRAWING SEC 100 44 WILLIE PEP   NOV 26, 20XX 14

INSTRUCTIONS:
Draw and dimension the assigned problems on A-size sheets at a scale where the grid is .20 inches (or 5 millimeters). You will need to vary the spacing between the views to provide room for the dimensions. It is recommended that you sketch the views on scratch paper in advance to determine the required space needed for your final solution. Dimension in either inches or millimeters as assigned. Lines and centerlines may be missing in all views.

The example at the left is an instrument-drawn working drawing which is more accurate than freehand sketched versions.

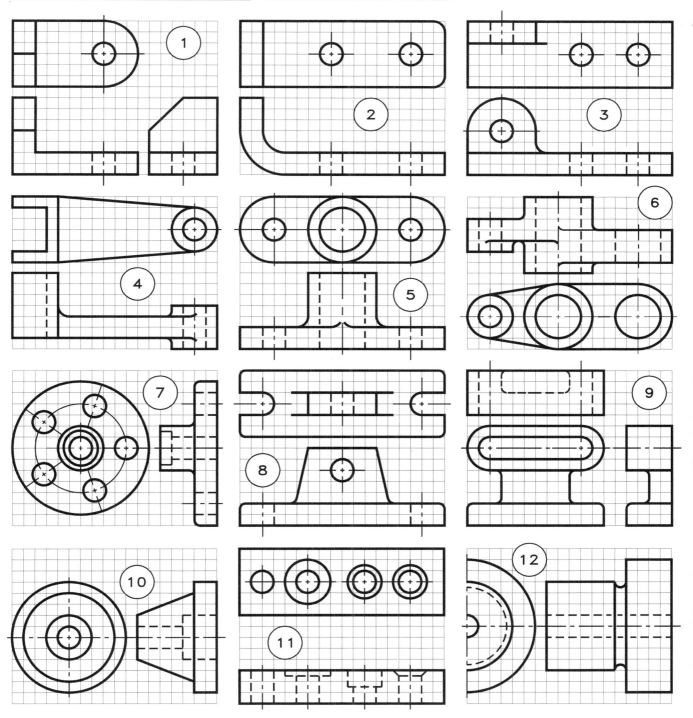

Graphics & Design
Dimensioning Applications

FILE: NAME:
SEC: DATE: MIN:

GRADE

128

A. Assembly of Mating Parts

The shaft must have a clearance fit inside the bushing.

The bushing must have a force fit inside the pulley.

2 BUSHING

3 SHAFT

1 PULLEY

B. Types of Cylindrical Fits

Force fit
2.004 / 2.000
1.997 / 1.994

Interference Fit

Force fit to clearance fit
2.006 / 2.000
2.002 / 1.998

Transition Fit

2.004 / 2.000
2.007 / 2.000
Same

Line Fit

C. Cylindrical Fits

Basic size=1.5000

Shaft tol.=.0025   Hole tol.=.0040

1.4950 / 1.4925   1.5040 / 1.5000

1. Limit Form

1.4925 / 1.4950   1.5040 / 1.5000

2. Meaning

Largest Shaft   Smallest Hole

1.4950   1.5000

1.5000 / −1.4950
+.0050 Allowance

3. Tightest Fit

Smallest Shaft   Largest Hole

1.4925   1.5040

1.5040 / −1.4925
+.0115 Max Clearance

4. Loosest Fit

D. An RC9 Clearance Fit: Table Data

Class RC9 (1.97-3.15 Range)

Limits of Clearance	(In thousandths) Hole	Shaft
9.0	7.0	−9.0
20.5	0	−13.5

HOLE: Basic Dia=2.5000

Upper Limit	Lower Limit
2.5000	2.5000
.0070	0
2.5070	2.5000

Ø 2.5070 / 2.5000

SHAFT: Basic Dia=2.5000

Upper Limit	Lower Limit
2.5000	2.5000
−.0090	−.0135
2.4910	2.4865

Ø 2.4910 / 2.4865

Limits of Clearance

2.5000	2.5070
2.4910	2.4865
+.0090	+.0205

Since basic DIA appears on hole, this is a Basic Hole System.

PROBLEM 1: By following Example D, complete the table of values for an RC5 clearance fit and a diameter of 3.0000 in the range of 1.97-3.15 found in ANSI table. These, and other tables can be found on the internet:

An RC 4 Clearance Fit:

Class RC5: Range = 1.97-3.15

Limits of Clearance	Hole	Shaft
1.2	1.8	−1.2
4.2	0	−2.4

(Thousandths of an Inch)

HOLE: Basic Dia=3.0000

Upper Limit   Lower Limit

SHAFT: Basic Dia=3.0000

Upper Limit   Lower Limit

Limits of Clearance

PROBLEM 2: Apply the tolerances found in Problem 1 above as dimensions to the hole and shaft below.

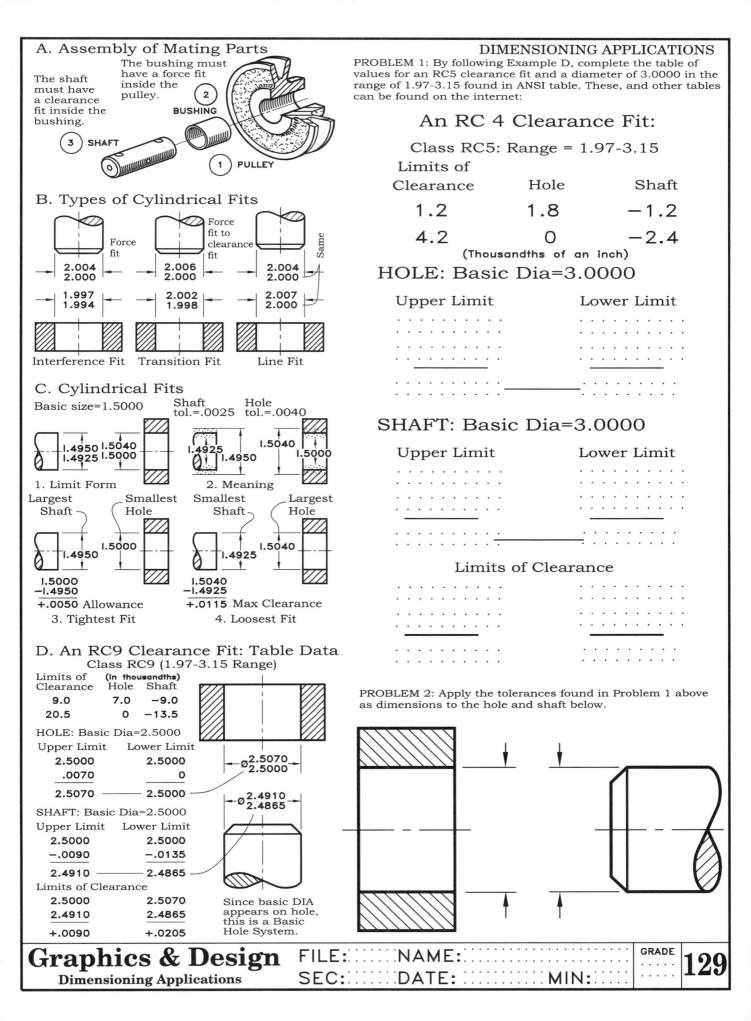

Graphics & Design
Dimensioning Applications

FILE: NAME:
SEC: DATE: MIN:

GRADE

129

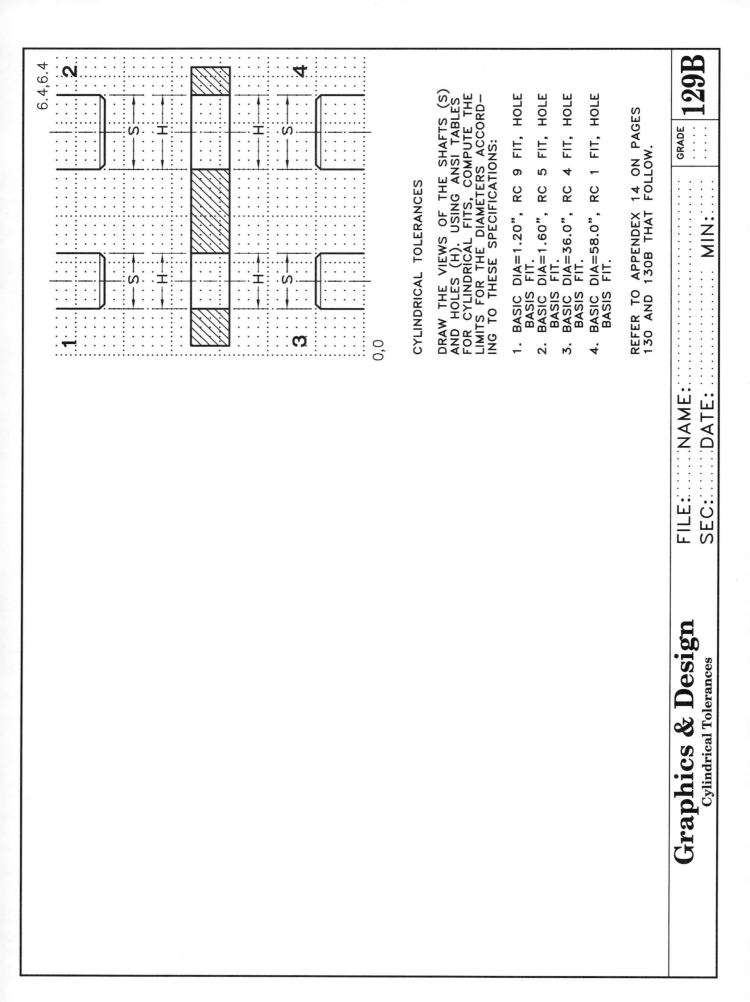

6.4,6.4

1   2

S

H

H

S

S

H

H

S

3   4

0,0

CYLINDRICAL TOLERANCES

DRAW THE VIEWS OF THE SHAFTS (S)
AND HOLES (H). USING ANSI TABLES
FOR CYLINDRICAL FITS, COMPUTE THE
LIMITS FOR THE DIAMETERS ACCORD-
ING TO THESE SPECIFICATIONS:

1. BASIC DIA=1.20", RC 9 FIT, HOLE
    BASIS FIT.
2. BASIC DIA=1.60", RC 5 FIT, HOLE
    BASIS FIT.
3. BASIC DIA=36.0", RC 4 FIT, HOLE
    BASIS FIT.
4. BASIC DIA=58.0", RC 1 FIT, HOLE
    BASIS FIT.

REFER TO APPENDEX 14 ON PAGES
130 AND 130B THAT FOLLOW.

Graphics & Design
Cylindrical Tolerances

FILE: .... NAME: ....
SEC: .... DATE: ....

GRADE 129B

.... MIN: ....

APPENDIX 14
AMERICAN STANDARD RUNNING AND SLIDING FITS (Hole Basis-Inches)

Limits are in thousandths of an inch.

Limits for hole and shaft are applied algebraically to the basic size to obtain the limits of size for the parts.

Data in boldface are in accordance with ABC agreements.

Symbols H5, g5, etc., are Hole and Shaft designations used in ABC System.

Nominal Size Range Inches Over	To	Class RC 1 Limits of Clearance	Standard Limits Hole H5	Shaft g4	Class RC 2 Limits of Clearance	Standard Limits Hole H6	Shaft g5	Class RC 3 Limits of Clearance	Standards Limits Hole H7	Shaft f6	Class RC 4 Limits of Clearance	Standard Limits Hole H8	Shaft f7
0	−0.12	0.1	+0.2	−0.1	0.1	+0.25	−0.1	0.3	+0.4	−0.3	0.3	+0.6	−0.3
		0.45	0	−0.25	0.55	0	−0.3	0.95	0	−0.55	1.3	0	−0.7
0.12	−0.24	0.15	+0.2	−0.15	0.15	+0.3	−0.15	0.4	+0.5	−0.4	0.4	+0.7	−0.4
		0.5	0	−0.3	0.65	0	−0.35	1.12	0	−0.7	1.6	0	−0.9
0.24	−0.40	0.2	0.25	−0.2	0.2	+0.4	−0.2	0.5	+0.6	−0.5	0.5	+0.9	−0.5
		0.6	0	−0.35	0.85	0	−0.45	1.5	0	−0.9	2.0	0	−1.1
0.40	−0.71	0.25	+0.3	−0.25	0.25	+0.4	−0.25	0.6	+0.7	−0.6	0.6	+1.0	−0.6
		0.75	0	−0.45	0.95	0	−0.55	1.7	0	−1.0	2.3	0	−1.3
0.71	−1.19	0.3	+0.4	−0.3	0.3	+0.5	−0.3	0.8	+0.8	−0.8	0.8	+1.2	−0.8
		0.95	0	−0.55	1.2	0	−0.7	2.1	0	−1.3	2.8	0	−1.6
1.19	−1.97	0.4	+0.4	−0.4	0.4	+0.6	−0.4	1.0	+1.0	−1.0	1.0	+1.6	−1.0
		1.1	0	−0.7	1.4	0	−0.8	2.6	0	−1.6	3.6	0	−2.0
1.97	−3.15	0.4	+0.5	−0.4	0.4	+0.7	−0.4	1.2	+1.2	−1.2	1.2	+1.8	−1.2
		1.2	0	−0.7	1.6	0	−0.9	3.1	0	−1.9	4.2	0	−2.4
3.15	−4.73	0.5	+0.6	−0.5	0.5	+0.9	−0.5	1.4	+1.4	−1.4	1.4	+2.2	−1.4
		1.5	0	−0.9	2.0	0	−1.1	3.7	0	−2.3	5.0	0	−2.8
4.73	−7.09	0.6	+0.7	−0.6	0.6	+1.0	−0.6	1.6	+1.6	−1.6	1.6	+2.5	−1.6
		1.8	0	−1.1	2.3	0	−1.3	4.2	0	−2.6	5.7	0	−3.2
7.09	−9.85	0.6	+0.8	−0.6	0.6	+1.2	−0.6	2.0	+1.8	−2.0	2.0	+2.8	−2.0
		2.0	0	−1.2	2.6	0	−1.4	5.0	0	−3.2	6.6	0	−3.8
9.85	−12.41	0.8	+0.9	−0.8	0.8	+1.2	−0.8	2.5	+2.0	−2.5	2.5	+3.0	−2.5
		2.3	0	−1.4	2.9	0	−1.7	5.7	0	−3.7	7.5	0	−4.5
12.41	−15.75	1.0	+1.0	−1.0	1.0	+1.4	−1.0	3.0	+	−3.0	3.0	+3.5	−3.0
		2.7	0	−1.7	3.4	0	−2.0	6.6	0	−4.4	8.7	0	−5.2
15.75	−19.69	1.2	+1.0	−1.2	1.2	+1.6	−1.2	4.0	+1.6	−4.0	4.0	+4.0	−4.0
		3.0	0	−2.0	3.8	0	−2.2	8.1	0	−5.6	10.5	0	−6.5
19.69	−30.09	1.6	+1.2	−1.6	1.6	+2.0	−1.6	5.0	+3.0	−5.0	5.0	+5.0	−5.0
		3.7	0	−2.5	4.8	0	−2.8	10.0	0	−7.0	13.0	0	−8.0
30.09	−41.49	2.0	+1.6	−2.0	2.0	+2.5	−2.0	6.0	+4.0	−6.0	6.0	+6.0	−6.0
		4.6	0	−3.0	6.1	0	−3.6	12.5	0	−8.5	16.0	0	−10.0
41.49	−56.19	2.5	+2.0	−2.5	2.5	+3.0	−2.5	8.0	+5.0	−8.0	8.0	+8.0	−8.0
		5.7	0	−3.7	7.5	0	−4.5	16.0	0	−11.0	21.0	0	−13.0
56.19	−76.39	3.0	+2.5	−3.0	3.0	+4.0	−3.0	10.0	+6.0	−10.0	10.0	+10.0	−10.0
		7.1	0	−4.6	9.5	0	−5.5	20.0	0	−14.0	26.0	0	−16.0
76.39	−100.9	4.0	+3.0	−4.0	4.0	+5.0	−4.0	12.0	+8.0	−12.0	12.0	+12.0	−12.0
		9.0	0	−6.0	12.0	0	−7.0	25.0	0	−17.0	32.0	0	−20.0
100.9	−131.9	5.0	+4.0	−5.0	5.0	+6.0	−5.0	16.0	+10.0	−16.0	16.0	+16.0	−16.0
		11.5	0	−7.5	15.0	0	−9.0	32.0	0	−22.0	36.0	0	−26.0
131.9	−171.9	6.0	+5.0	−6.0	6.0	+8.0	−6.0	18.0	+8.0	−18.0	18.0	+20.0	−18.0
		14.0	0	−9.0	19.0	0	−11.0	38.0	0	−26.0	50.0	0	−30.0
171.9	−200	8.0	+6.0	−8.0	8.0	+10.0	−8.0	22.0	+16.0	−22.0	22.0	+25.0	−22.0
		18.0	0	−12.0	22.0	0	−12.0	48.0	0	−32.0	63.0	0	−38.0

Source: Courtesy of USASI; B4.1—1955.

APPENDIX 14 (Continued)
AMERICAN STANDARD RUNNING AND SLIDING FITS (Hole Basis-Inches)

Class RC 5			Class RC 6			Class RC 7			Class RC 8			Class RC 9			Nominal Size Range Inches	
Limits of Clearance	Hole H8	Shaft e7	Limits of Clearance	Hole H9	Shaft e8	Limits of Clearance	Hole H9	Shaft d8	Limits of Clearance	Hole H10	Shaft c9	Limits of Clearance	Hole H11	Shaft	Over	To
0.6	+0.6	−0.6	0.6	+1.0	−0.6	1.0	+1.0	−1.0	2.5	+1.6	−2.5	4.0	+2.5	−4.0	0	0.12
1.6	−0	−1.0	2.2	−0	−1.2	2.6	0	−1.6	5.1	0	−3.5	8.1	0	−5.6		
0.8	+0.7	−0.8	0.8	+1.2	−0.8	1.2	+1.2	−1.2	2.8	+1.8	−2.8	4.5	+3.0	−4.5	0.12	0.24
2.0	−0	−1.3	2.7	−0	−1.5	3.1	0	−1.9	5.8	0	−4.0	9.0	0	−6.0		
1.0	+0.9	−1.0	1.0	+1.4	−1.0	1.6	+1.4	−1.6	3.0	+2.2	−3.0	5.0	+3.5	−5.0	0.24	0.40
2.5	−0	−1.16	3.3	−0	−1.9	3.9	0	−2.5	6.6	0	−4.4	10.7	0	−7.2		
1.2	+1.0	−1.2	1.2	+1.6	−1.2	2.0	+1.6	−2.0	3.5	+2.8	−3.5	6.0	+4.0	−6.0	0.40	0.71
2.9	−0	−1.9	3.8	−0	−2.2	4.6	0	−3.0	7.9	0	−5.1	12.8	−0	−8.8		
1.6	+1.2	−1.6	1.6	+2.0	−1.6	2.5	+2.0	−2.5	4.5	+3.5	−4.5	7.0	+5.0	−7.0	0.71	1.19
3.6	−0	−2.4	4.8	−0	−2.8	5.7	0	−3.7	10.0	0	−6.5	15.5	0	−10.5		
2.0	+1.6	−2.0	2.0	+2.5	−2.0	3.0	+2.5	−3.0	5.0	+4.0	−5.0	8.0	+6.0	−8.0	1.19	1.97
4.6	−0	−3.0	6.1	−0	−3.6	7.1	0	−4.6	11.5	0	−7.5	18.0	0	−12.0		
2.5	+1.8	−2.5	2.5	+3.0	−2.5	4.0	+3.0	−4.0	6.0	+4.5	−6.0	9.0	+7.0	−9.0	1.97	3.15
5.5	−0	−3.7	7.3	−0	−4.3	8.8	0	−5.8	13.5	0	−9.0	20.5	0	−13.5		
3.0	+2.2	−3.0	3.0	+3.5	−3.0	5.0	+3.5	−5.0	7.0	+5.0	−7.0	10.0	+9.0	−10.0	3.15	4.73
6.6	−0	−4.4	8.7	−0	−5.2	10.7	0	−7.2	15.5	0	−10.5	24.0	0	−15.0		
3.5	+2.5	−3.5	3.5	+4.0	−3.5	6.0	+4.0	−6.0	8.0	+6.0	−8.0	12.0	+10.0	−12.0	4.73	7.09
7.6	−0	−5.1	10.0	−0	−6.0	12.5	0	−8.5	18.0	0	−12.0	28.0	0	−18.0		
4.0	+2.8	−4.0	4.0	+4.5	−4.0	7.0	+4.5	−7.0	10.0	+7.0	−10.0	15.0	+12.0	−15.0	7.09	9.85
8.6	−0	−5.8	11.3	0	−6.8	14.3	0	−9.8	21.5	0	−14.5	34.0	0	−22.0		
5.0	+3.0	−5.0	5.0	+5.0	−5.0	8.0	+5.0	−8.0	12.0	+8.0	−12.0	18.0	+12.0	−18.0	9.85	12.41
10.0	0	−7.0	13.0	0	−8.0	16.0	0	−11.0	25.0	0	−17.0	38.0	0	−26.0		
6.0	+3.5	−6.0	6.0	+6.0	−6.0	10.0	+6.0	−10.0	14.0	+9.0	−14.0	22.0	+14.0	−22.0	12.41	15.75
11.7	0	−8.2	15.5	0	−9.5	19.5	0	13.5	29.0	0	−20.0	45.0	0	−31.0		
8.0	+4.0	−8.0	8.0	+6.0	−8.0	12.0	+6.0	−12.0	16.0	+10.0	−16.0	25.0	+16.0	−25.0	15.75	19.69
14.5	0	−10.5	18.0	0	−12.0	22.0	0	−16.0	32.0	0	−22.0	51.0	0	−35.0		
10.0	+5.0	−10.0	10.0	+8.0	−10.0	16.0	+8.0	−16.0	20.0	+12.0	−20.0	30.0	+20.0	−30.0	19.69	30.09
18.0	0	−13.0	23.0	0	−15.0	29.0	0	−21.0	40.0	0	−28.0	62.0	0	−42.0		
12.0	+6.0	−12.0	12.0	+10.0	−12.0	20.0	+10.0	−20.0	25.0	+16.0	−25.0	40.0	+25.0	−40.0	30.09	41.49
22.0	0	−16.0	28.0	0	−18.0	36.0	0	−26.0	51.0	0	−35.0	81.0	0	−56.0		
16.0	+8.0	−16.0	16.0	+12.0	−16.0	25.0	+12.0	−25.0	30.0	+20.0	−30.0	50.0	+30.0	−50.0	41.49	56.19
29.0	0	−21.0	36.0	0	−24.0	45.0	0	−33.0	62.0	0	−42.0	100	0	−70.0		
20.0	+10.0	−20.0	20.0	+16.0	−20.0	30.0	+16.0	−30.0	40.0	+25.0	−40.0	60.0	+40.0	−60.0	56.19	76.39
36.0	0	−26.0	46.0	0	−30.0	56.0	0	−40.0	81.0	0	−56.0	125	0	−85.0		
25.0	+12.0	−25.0	25.0	+20.0	−25.0	40.0	+20.0	−40.0	50.0	+30.0	−50.0	80.0	+50.0	−80.0	76.39	100.9
45.0	0	−33.0	57.0	0	−37.0	72.0	0	−52.0	100	0	−70.0	160	0	−110		
30.0	+16.0	−30.0	30.0	+35.0	−30.0	50.0	+25.0	−50.0	60.0	+40.0	−60.0	100	+60.0	−100	100.9	131.9
56.0	0	−40.0	71.0	0	−46.0	91.0	0	−66.0	125	0	−85.0	200	0	−140		
35.0	+20.0	−35.0	35.0	+30.0	−35.0	60.0	+30.0	−60.0	80.0	+50.0	−80.0	130	+80.0	−130	131.9	171.9
57.0	0	−47.0	85.0	0	−55.0	110.0	0	−80.0	160	0	−110	260	0	−180		
45.0	+25.0	−45.0	45.0	+40.0	−45.0	80.0	+40.0	−80.0	100	+60.0	−100	150	+100	−150	171.9	200
86.0	0	−61.0	110.0	0	−70.0	145.0	0	−105.0	200	0	−140	310	0	−210		

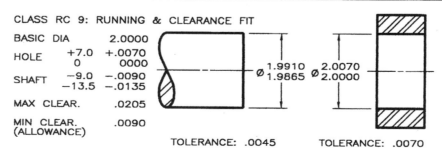

CLASS RC 9: RUNNING & CLEARANCE FIT

BASIC DIA		2.0000
HOLE	+7.0 / 0	+.0070 / 0000
SHAFT	−9.0 / −13.5	−.0090 / −.0135
MAX CLEAR.		.0205
MIN CLEAR. (ALLOWANCE)		.0090

Ø 1.9910 / 1.9865   Ø 2.0070 / 2.0000

TOLERANCE: .0045   TOLERANCE: .0070

A. Preferred Basic Sizes (Millimeters)

First Choice	Second Choice	First Choice	Second Choice	First Choice	Second Choice
1		10		100	
	1.1		11		110
1.2		12		120	
	1.4		14		140
1.6		16		160	
	1.8		18		180
2		20		200	
	2.2		22		220
2.5		25		250	
	2.8		28		280
3		30		300	
	3.5		35		350
4		40		400	
	4.5		45		450
5		50		500	
	5.5		55		550
6		60		600	
	7		70		700
8		80		800	
	9		90		900
				1000	

PROB. 1: By following Example C, complete the table of values for a D9/h9 close running fit and a diameter of 80 in the shaft basis, free running clearance fit (ANSI B4.2). This table can be found on Page 132 or on the internet along with a dozen other tables for metric fits.

Metric Fit
Close Running Fit: D9/h9
Basic size=80 mm
(From American National Standard Tables: ANSI B4.2 available on the Internet)

B. Preferred Fits: Metric System

Hole Basis	Shaft Basis		Description
H11/c11	C11/h11	Clearance	**Loose Running Fit** for wide commerical tolerances on external members.
H9/d9	D9/h9	Clearance	**Free Running Fit** for large temperature variations, high running speeds, or high journal pressures.
H8/f7	F8/h7	Clearance	**Close Running Fit** for accurate location and moderate speeds and journal pressures.
H7/g6	G7/h6	Clearance	**Sliding Fit** for accurate fit and location and free moving and turning, not free running.
H7/h6	H7/h6	Transition	**Locational Clearance** for snug fits for parts that can be freely assembled.
H7/k6	K7/h6	Transition	**Locational Transition Fit** for accurate locations.
H7/n6	N7/h6	Transition	**Locational Transition Fit** for more accurate locations and greater interference.
H7/p6	P7/h6	Interference	**Locational Interference Fit** for rigidity and alignment without special bore pressures.
H7/s6	S7/h6	Interference	**Medium Drive Fit** for shrink fits on light sections; tightest fit usable for cast iron.
H7/u6	U7/h6	Interference	**Force Fit** for parts that can be highly stressed and for shrink fits.

Close Running Fit

Hole H8	Shaft f7	Fit
. . . .	. . . .	. . . .
. . . .	. . . .	. . . .
. . . .	. . . .	. . . .

Shaft tolerance = ...

Hole tolerance = ...

Tightest Fit	Loosest Fit
. . . .	. . . .

	Upper Deviation	Lower Deviation
Shaft	. . . .	. . . .
Hole	. . . .	. . . .

C. Close Running Fit: H8/f7

Basic size=50 mm
(From Appendix table)

Close Running Fit

Hole H8	Shaft f7	FIT
50.039	49.975	0.089
50.000	49.950	0.025

Shaft tolerance = 0.025

Hole tolerance = 0.039

Tightest Fit	Loosest Fit
0.025	0.089

	Upper Deviation	Lower Deviation
Shaft	−0.025	−0.050
Hole	+0.039	0.000

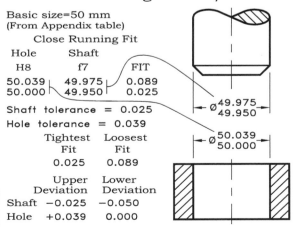

Ø 49.975 / 49.950

Ø 50.039 / 50.000

This drawing illustrates how calculations are made and applied to a hole and shaft using data taken from the ANSI B4.2 metric fit tables.

PROBLEM 2: Apply the tolerances found in Problem 1 above as dimensions to the hole and shaft below.

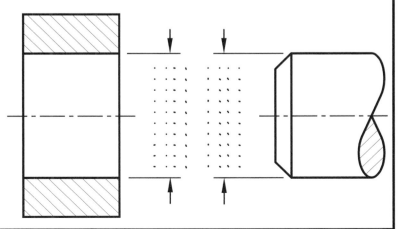

APPENDIX 15
SHAFT-BASIS CLEARANCE FITS-CYLINDRICAL FITS (Metric)

Basic Size		Loose Running			Free Running			Close Running			Sliding			Locational Clearance		
		Hole C11	Shaft h11	Fit	Hole D9	Shaft h9	Fit	Hole F8	Shaft h7	Fit	Hole G7	Shaft h6	Fit	Hole H7	Shaft h6	Fit
1	MAX	1.120	1.000	0.180	1.045	1.000	0.070	1.020	1.000	0.030	1.012	1.000	0.018	1.010	1.000	0.016
	MIN	1.060	0.940	0.060	1.020	0.975	0.020	1.006	0.990	0.006	1.002	0.994	0.002	1.000	0.994	0.000
1.2	MAX	1.320	1.200	0.180	1.245	1.200	0.070	1.220	1.200	0.030	1.212	1.200	0.018	1.210	1.200	0.016
	MIN	1.260	1.140	0.060	1.220	1.175	0.020	1.206	1.190	0.006	1.202	1.194	0.002	1.200	1.194	0.000
1.6	MAX	1.720	1.600	0.180	1.656	1.600	0.070	1.620	1.600	0.030	1.612	1.600	0.018	1.610	1.600	0.016
	MIN	1.660	1.540	0.060	1.620	1.575	0.020	1.606	1.590	0.006	1.602	1.595	0.002	1.600	1.594	0.000
2	MAX	2.120	2.000	0.180	2.045	2.000	0.070	2.020	2.000	0.030	2.012	2.000	0.018	2.010	2.000	0.016
	MIN	2.060	1.940	0.060	2.020	1.975	0.020	2.006	1.990	0.006	2.002	1.994	0.002	2.000	1.994	0.000
2.5	MAX	2.620	2.500	0.180	2.545	2.500	0.070	2.520	2.500	0.030	2.512	2.500	0.018	2.510	2.500	0.016
	MIN	2.560	2.440	0.060	2.520	2.475	0.020	2.506	2.490	0.006	2.502	2.494	0.002	2.500	2.494	0.000
3	MAX	3.120	3.000	0.180	3.045	3.000	0.070	3.020	3.000	0.030	3.012	3.000	0.018	3.010	3.000	0.016
	MIN	3.060	2.940	0.060	3.020	2.975	0.020	3.006	2.990	0.006	3.002	2.994	0.002	3.000	2.994	0.000
4	MAX	4.145	4.000	0.220	4.060	4.000	0.090	4.028	4.000	0.040	4.016	4.000	0.024	4.012	4.000	0.020
	MIN	4.070	3.925	0.070	4.030	3.970	0.030	4.010	3.988	0.010	4.004	3.992	0.004	4.000	3.992	0.000
5	MAX	5.145	5.000	0.220	5.060	5.000	0.090	5.028	5.000	0.040	5.016	5.000	0.024	5.012	5.000	0.020
	MIN	5.070	4.925	0.070	5.030	4.970	0.030	5.010	4.988	0.010	5.004	4.992	0.004	5.000	4.992	0.000
6	MAX	6.145	6.000	0.220	6.060	6.000	0.090	6.028	6.000	0.040	6.016	6.000	0.024	6.012	6.000	0.020
	MIN	6.070	5.925	0.070	6.030	5.970	0.030	6.010	5.988	0.010	6.004	5.992	0.004	6.000	5.992	0.000
8	MAX	8.170	8.000	0.260	8.076	8.000	0.112	8.035	8.000	0.050	8.020	8.000	0.029	8.015	8.000	0.024
	MIN	8.080	7.910	0.080	8.040	7.964	0.040	8.013	7.985	0.013	8.005	7.991	0.005	8.000	7.991	0.000
10	MAX	10.170	10.000	0.260	10.076	10.000	0.112	10.035	10.000	0.050	10.020	10.000	0.029	10.015	10.000	0.024
	MIN	10.080	9.910	0.080	10.040	9.964	0.040	10.013	9.985	0.013	10.005	9.991	0.005	10.000	9.991	0.000
12	MAX	12.205	12.000	0.315	12.093	12.000	0.136	12.043	12.000	0.061	12.024	12.000	0.035	12.018	12.000	0.029
	MIN	12.095	11.890	0.095	12.050	11.957	0.050	12.016	11.982	0.016	12.006	11.989	0.006	12.000	11.989	0.000
16	MAX	16.205	16.000	0.315	16.093	16.000	0.136	16.043	16.000	0.061	16.024	16.000	0.035	16.018	16.000	0.029
	MIN	16.095	15.890	0.095	16.050	15.957	0.050	16.016	15.982	0.016	16.006	15.989	0.006	16.000	15.989	0.000
20	MAX	20.240	20.000	0.370	20.117	20.000	0.169	20.053	20.000	0.074	20.028	20.000	0.041	20.021	20.000	0.034
	MIN	20.110	19.870	0.110	20.065	19.948	0.065	20.020	19.979	0.020	20.007	19.987	0.007	20.000	19.987	0.000
25	MAX	25.240	25.000	0.370	25.117	25.000	0.169	25.053	25.000	0.074	25.028	25.000	0.041	25.021	25.000	0.034
	MIN	25.110	24.870	0.110	25.065	24.948	0.065	25.020	24.979	0.020	25.007	24.987	0.007	25.000	24.987	0.000
30	MAX	30.240	30.000	0.370	30.117	30.000	0.169	30.053	30.000	0.074	30.028	30.000	0.041	30.021	30.000	0.034
	MIN	30.110	29.870	0.110	30.065	29.948	0.065	30.020	29.979	0.020	30.007	29.987	0.007	30.000	29.987	0.000

Source: American National Standard Preferred Metric Limits and Fits, ANSI B4.2. Dimensions are in mm.

APPENDIX 15
SHAFT-BASIS CLEARANCE FITS-CYLINDRICAL FITS (Metric)

Basic Size		Loose Running			Free Running			Close Running			Sliding			Locational Clearance		
		Hole C11	Shaft h11	Fit	Hole D9	Shaft h9	Fit	Hole F8	Shaft h7	Fit	Hole G7	Shaft h6	Fit	Hole H7	Shaft h6	Fit
40	MAX	40.280	40.000	0.440	40.142	40.000	0.204	40.064	40.000	0.089	40.034	40.000	0.050	40.025	40.000	0.041
	MIN	40.120	39.840	0.120	40.080	39.938	0.080	40.025	39.975	0.025	40.009	39.984	0.009	40.000	39.984	0.000
50	MAX	50.290	50.000	0.450	50.142	50.000	0.204	50.064	50.000	0.089	50.034	50.000	0.050	50.025	50.000	0.041
	MIN	50.130	49.840	0.130	50.080	49.938	0.080	50.025	49.975	0.025	50.009	49.984	0.009	50.000	49.984	0.000
60	MAX	60.330	60.000	0.520	60.174	60.000	0.248	60.076	60.000	0.106	60.040	60.000	0.059	60.030	60.000	0.049
	MIN	60.140	59.810	0.140	60.100	59.926	0.100	60.030	59.970	0.030	60.010	59.981	0.010	60.000	59.981	0.000
80	MAX	80.340	80.000	0.530	80.174	80.000	0.248	80.076	80.000	0.106	80.040	80.000	0.059	80.030	80.000	0.049
	MIN	80.150	79.810	0.150	80.100	79.926	0.100	80.030	79.970	0.030	80.010	79.981	0.010	80.000	79.981	0.000
100	MAX	100.390	100.000	0.610	100.207	100.000	0.294	100.090	100.000	0.125	100.047	100.000	0.069	100.035	100.000	0.057
	MIN	100.170	99.780	0.170	100.120	99.913	0.120	100.036	99.965	0.036	100.012	99.979	0.012	100.000	99.979	0.000
120	MAX	120.400	120.000	0.620	120.207	120.000	0.294	120.090	120.000	0.125	120.047	120.000	0.069	120.035	120.000	0.057
	MIN	120.180	119.780	0.180	120.120	119.913	0.120	120.036	119.965	0.036	120.012	119.978	0.012	120.000	119.978	0.000
160	MAX	160.460	160.000	0.710	160.245	160.000	0.345	160.106	160.000	0.146	160.054	160.000	0.079	160.040	160.000	0.065
	MIN	160.210	159.750	0.210	160.145	159.900	0.145	160.043	159.960	0.043	160.014	159.975	0.014	160.000	159.975	0.000
200	MAX	200.530	200.000	0.820	200.285	200.000	0.400	200.122	200.000	0.168	200.061	200.000	0.090	200.046	200.000	0.075
	MIN	200.240	199.710	0.240	200.170	199.885	0.170	200.050	199.954	0.050	200.015	199.971	0.015	200.000	199.971	0.000
250	MAX	250.570	250.000	0.860	250.285	250.000	0.400	250.122	250.000	0.168	250.061	250.000	0.090	250.046	250.000	0.075
	MIN	250.280	249.710	0.280	250.170	249.885	0.170	250.050	249.954	0.050	250.015	249.971	0.015	250.000	249.971	0.000
300	MAX	300.650	300.000	0.970	300.320	300.000	0.450	300.137	300.000	0.189	300.069	300.000	0.101	300.052	300.000	0.084
	MIN	300.330	299.680	0.330	300.190	299.870	0.190	300.056	299.948	0.056	300.017	299.968	0.017	300.000	299.968	0.0'
400	MAX	400.760	400.000	1.120	400.350	400.000	0.490	400.151	400.000	0.208	400.075	400.000	0.111	400.057	400.000	0.9.
	MIN	400.400	399.640	0.400	400.210	399.860	0.210	400.062	399.943	0.062	400.018	399.964	0.018	400.000	399.964	0.000
500	MAX	500.880	500.000	1.280	500.385	500.000	0.540	500.165	500.000	0.228	500.083	500.000	0.123	500.063	500.000	0.103
	MIN	500.480	499.600	0.480	500.230	499.845	0.230	500.068	499.937	0.068	500.020	499.960	0.020	500.000	499.960	0.000

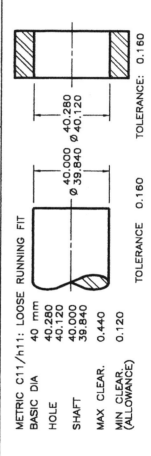

METRIC C11/h11: LOOSE RUNNING FIT

BASIC DIA   40 mm

HOLE   40.280 / 40.120

SHAFT   40.000 / 39.840

MAX CLEAR.   0.440

MIN CLEAR. (ALLOWANCE) 0.120

Ø 40.000 / Ø 39.840   TOLERANCE 0.160

Ø 40.280 / Ø 40.120   TOLERANCE: 0.160

A. Flatness Tolerance

PROBLEM. 1: By following Examples A & B, note the part to have a Flatness of 0.50 and the cylinder to have a Straightness of 0.2. Overall size dimensions may be omitted for all problem here.

⌖ 0.40

0.40 Tolerance

1. Drawing   2. Meaning

1.

Flatness
Straightness

B. Staightness Tolerance

— 0.12

0.12 Tolerance

Ø15.50 / 15.40
Ø15.62

1. Drawing Notes   2. Meaning

C. Circularity (Roundness)

○ 0.54   ○ 0.54

1. Drawing: Cylinder   2. Drawing: Cone

PROBLEM 2: By following Example C, note the conical and cylindrical features to have a Roundness tolerance of 0.86. Overall size dimensions may be omitted.

0.54 Tolerance zone

SECTION

Meaning: Cylinder   Meaning: Cone

2.

D. Cylindricity Tolerance

Ø50.00−50.75

*Cylindricity tolerance must be less than size tolerance

⌭ 0.54

0.54 Tolerance Zone

1. Drawing   2. Meaning

PROBLEM 3: By following Example D, note the part to have a Cylindricity of 0.74, measure and dimension the part's diameter, measured as full size.

3.

E. Profile Tolerance

B

⌒ 0.2 A B

R45

R45

XX

A

90

1. Drawing

0.2 Bilateral Tolerance Zone

Datum B
Datum A

BASIC

2. Meaning

Graphics & Design
Tolerances of Form: Metric

FILE: NAME:
SEC: DATE: MIN:

GRADE

133

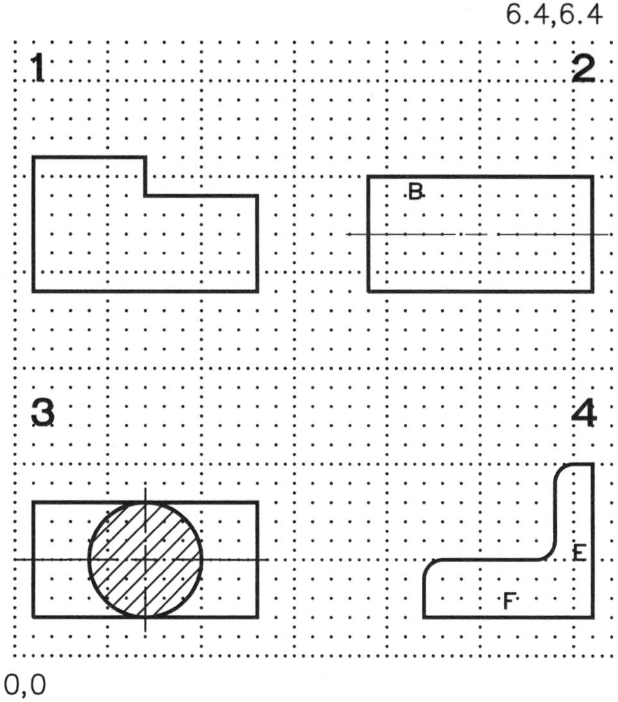

6.4,6.4

1

2

3

4

B

E

F

0,0

TOLERANCES OF FORM

USING SYMBOLS, NOTE THE FOLLOWING FEATURES OF THE PARTS ABOVE:

1. SURFACE A IS FLAT WITH 0.40 mm.
2. CYLINDER B IS STRAIGHT WITHIN 0.30 mm.
3. THE CYLINDER IS ROUND WITHIN 0.20.
4. SURFACE E IS PERPENDICULAR TO DATUM F WITHIN 0.30 mm.

Graphics & Design
Tolerances of Form

FILE:........... NAME:..........
SEC:........... DATE:..........

GRADE 133B

MIN:......

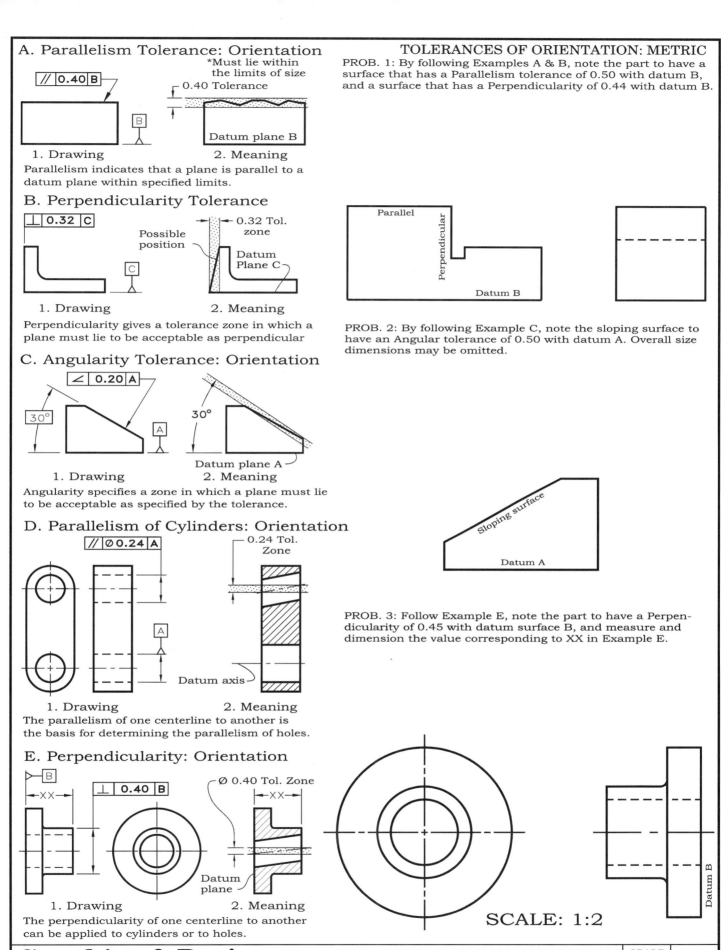

A. Parallelism Tolerance: Orientation

Must lie within the limits of size

// 0.40 B

0.40 Tolerance

Datum plane B

1. Drawing   2. Meaning

Parallelism indicates that a plane is parallel to a datum plane within specified limits.

B. Perpendicularity Tolerance

⊥ 0.32 C

0.32 Tol. zone

Possible position

Datum Plane C

1. Drawing   2. Meaning

Perpendicularity gives a tolerance zone in which a plane must lie to be acceptable as perpendicular

C. Angularity Tolerance: Orientation

∠ 0.20 A

30°

30°

Datum plane A

1. Drawing   2. Meaning

Angularity specifies a zone in which a plane must lie to be acceptable as specified by the tolerance.

D. Parallelism of Cylinders: Orientation

// Ø 0.24 A

0.24 Tol. Zone

A

Datum axis

1. Drawing   2. Meaning

The parallelism of one centerline to another is the basis for determining the parallelism of holes.

E. Perpendicularity: Orientation

B

⊥ 0.40 B

Ø 0.40 Tol. Zone

XX

XX

Datum plane

1. Drawing   2. Meaning

The perpendicularity of one centerline to another can be applied to cylinders or to holes.

TOLERANCES OF ORIENTATION: METRIC

PROB. 1: By following Examples A & B, note the part to have a surface that has a Parallelism tolerance of 0.50 with datum B, and a surface that has a Perpendicularity of 0.44 with datum B.

Parallel

Perpendicular

Datum B

PROB. 2: By following Example C, note the sloping surface to have an Angular tolerance of 0.50 with datum A. Overall size dimensions may be omitted.

Sloping surface

Datum A

PROB. 3: Follow Example E, note the part to have a Perpendicularity of 0.45 with datum surface B, and measure and dimension the value corresponding to XX in Example E.

SCALE: 1:2

Datum B

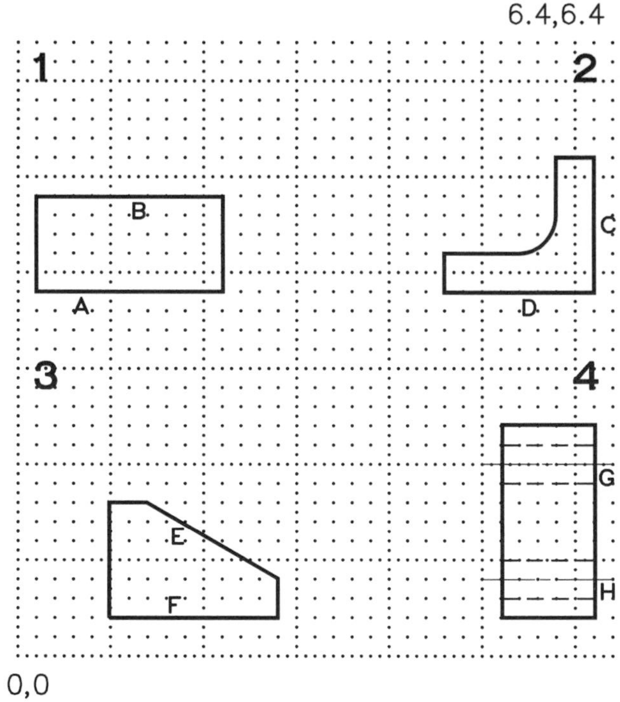

6.4, 6.4

1

2

3

4

B

A

C

D

E

F

G

H

0,0

TOLERANCES OF FORM

USING SYMBOLS, NOTE THE FOLLOWING
FEATURES OF THE PARTS ABOVE.

1. SURFACE A IS PARALLEL TO DATUM
   SURFACE B WITHIN 0.30 mm.
2. SURFACE C IS PERPENDICULAR TO
   DATUM SURFACE D WITHIN 0.26 mm.
3. SURFACE E HAS AN ANGULARITY
   WITHIN 0.60 OF DATUM F.
4. THE HOLE AT G IS PARALLEL TO THE
   HOLE AT H WITHIN 0.55 mm.

Graphics & Design
Tolerances of Form

FILE: NAME:
SEC: DATE:

GRADE 134B MIN:

A. The purpose of the working drawing is to bring concepts into reality and to show how they are made and assembled.

DRAWING SHEET SIZES

ENGLISH SIZES			METRIC SIZES		
A	11 X 8.5		A4	297 X 210
B	17 X 11		A3	420 X 297
C	22 X 17		A2	594 X 420
D	34 X 22		A1	841 X 594
E	44 X 34		A0	1189 X 841

B. The standard sizes for working drawing sheets are shown here in English (inches) and Metric (millimeters) units.

INSTRUCTIONS: Draw the title block from C above for the part, Shaft Support, placed against the corner with the parts list placed above. Use 1/8 letters and list the following parts: 1 BASE, 1 REQ., C.I.; 2 SHAFT, 1 REQ., 1020 STEEL; and 3 FORK, 1 REQ., 1020 STEEL.

2	SHAFT	2	1020 STL
1	BASE	1	CAST IRON
NO	PART NAME	REQ	MATERIAL
	PARTS LIST		

.38

◄———— 5" (Approximately) ————►

TITLE BLOCK	
BY: **RED GRANGE**	SECT: **500**
DATE: **MAY 2, 2XXX**	SHEET 1
SCALE: **FULL SIZE**	OF 1 SHEETS

.38

C. These blocks are adequate for classroom assignments, but those used in industry are usually detailed as they are legal documents. Those above are partially printed.

REVISIONS	COMPANY NAME Address		
CHG. HEIGHT	TITLE: **LEFT—END BEARING**		
FAO	DRAWN BY: **JOHNNY RINGO**		
	CHECKED BY: **FRED J. DODGE**		
	DATE: **JULY 14, XXXX**		
	SCALE: **HALF SIZE**	SHEET 1 OF 3 SHEETS

D. This title block has a "Revisions" block added for the addition of recommended changes in the details of the parts.

E. The System Internatial symbol indicates that metric units were used and the truncated cone indicates it is a third-angle projection.

RIGHT-BORDER OF THE SHEET

LOWER BORDER OF SHEET

SI ▷ ⊕

Graphics & Design

Assignment Solution

FILE: .... NAME: ....

SEC: .... DATE: ....

① PULLEY
1020 STEEL
1 REQUIRED

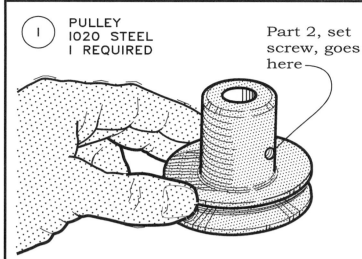

As the saying goes, "A picture is worth a thousand words."

Part 2, set screw, goes here

INSTRUCTIONS:
A. The A-pulley is to be drawn on sheet 137 as shown below, using the principles that you have learned so far.
B. You can see from the example sheet that it is drawn and dimensioned in millimeters.
C. This assignment is indeed "copy work," but this is the most efficient way to become familiar with the steps of making a working drawing.
D. When finished, replace the name, "Wyatt S. Earp" with your name preceded by your file number, "12 JOHN DOE" for example.
E. As always, strive for good line work and good lettering to the best of your ability. And, look for errors that should be corrected or improved.

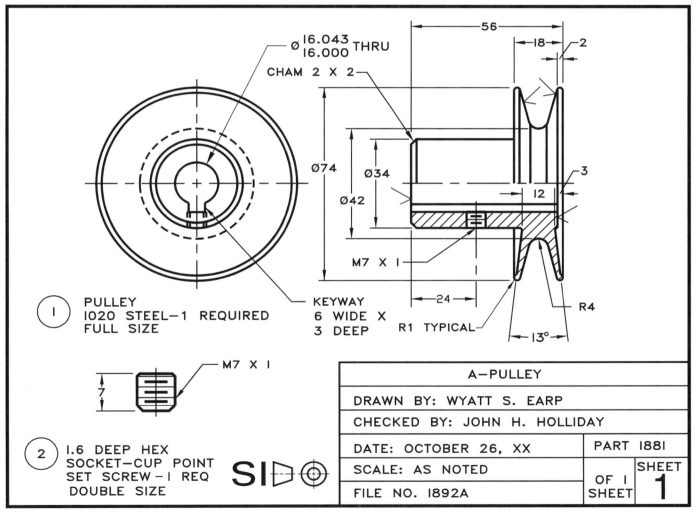

① PULLEY
1020 STEEL–1 REQUIRED
FULL SIZE

Ø 16.043 / 16.000 THRU
CHAM 2 X 2

KEYWAY
6 WIDE X
3 DEEP

M7 X 1
R1 TYPICAL

56
18 — 2
Ø74
Ø34
Ø42
3
12
R4
24
13°

M7 X 1
7

② 1.6 DEEP HEX
SOCKET–CUP POINT
SET SCREW–1 REQ
DOUBLE SIZE

SI

A–PULLEY		
DRAWN BY: WYATT S. EARP		
CHECKED BY: JOHN H. HOLLIDAY		
DATE: OCTOBER 26, XX	PART 1881	
SCALE: AS NOTED	OF 1 SHEET	SHEET 1
FILE NO. 1892A		

An A-size sheet is 8.5" x 11" with a border placed about 10 mm on all sides within the overall sheet size.

Graphics & Design
Working Drawings

FILE:⋯⋯ NAME:⋯⋯⋯⋯⋯
SEC:⋯⋯ DATE:⋯⋯⋯ MIN:⋯⋯

GRADE
136

6.4,6.4

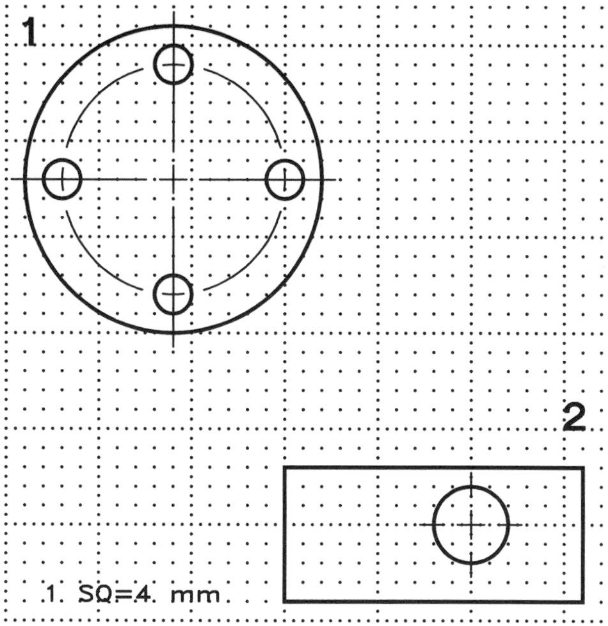

1

2

.1 SQ=4. mm

0,0

GEOMETRIC TOLERANCING

1. THE HOLES ARE AT TRUE POSITION
   WITHIN 0.08 mm DIA AND EACH
   HOLE HAS A TOLERANCE OF 0.4 mm.
2. THE HOLE IS LOCATED WITHIN A
   0.60 mm DIA AND IT HAS A SIZE
   TOLERANCE OF 0.4 mm.

Graphics & Design
Geometric Tolerancing: Metric

FILE: .... NAME: ....
SEC: .... DATE: ....

GRADE **136B**

MIN: ....

WORKING DRAWING OF PAGE 136

SHEET
1

OF 1 SHEET

SI

Graphics & Design

Assignment Solution

FILE: NAME:

SEC: DATE: MIN:

GRADE

A. Pictorial Assembly: Assembled

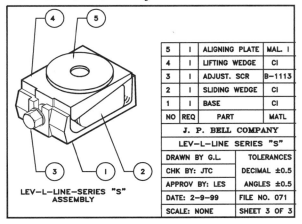

5	1	ALIGNING PLATE	MAL. 1
4	1	LIFTING WEDGE	CI
3	1	ADJUST. SCR	B-1113
2	1	SLIDING WEDGE	CI
1	1	BASE	CI
NO	REQ	PART	MATL

J. P. BELL COMPANY	
LEV-L-LINE SERIES "S"	
DRAWN BY G.L.	TOLERANCES
CHK BY: JTC	DECIMAL ±0.5
APPROV BY: LES	ANGLES ±0.5
DATE: 2-9-99	FILE NO. 071
SCALE: NONE	SHEET 3 OF 3

LEV-L-LINE-SERIES "S"
ASSEMBLY

B. Exploded Orthographic Assembly

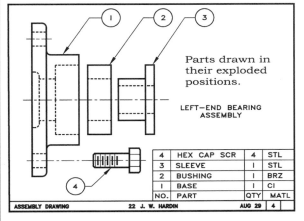

Parts drawn in
their exploded
positions.

LEFT-END BEARING
ASSEMBLY

4	HEX CAP SCR	4	STL
3	SLEEVE	1	STL
2	BUSHING	1	BRZ
1	BASE	1	CI
NO.	PART	QTY	MATL

ASSEMBLY DRAWING   22 J. W. HARDIN   AUG 29 | 4

C. Assembled Orthographic Section

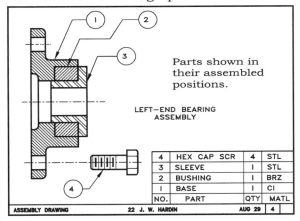

Parts shown in
their assembled
positions.

LEFT-END BEARING
ASSEMBLY

4	HEX CAP SCR	4	STL
3	SLEEVE	1	STL
2	BUSHING	1	BRZ
1	BASE	1	CI
NO.	PART	QTY	MATL

ASSEMBLY DRAWING   22 J. W. HARDIN   AUG 29 | 4

D. Exploded, Sectioned, Pictorial Assembly

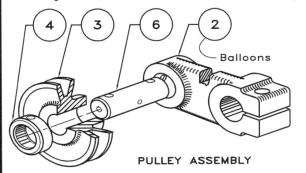

Balloons

PULLEY ASSEMBLY

HANGER-ROD SUPPORT

Develop the preliminary sketches of the hanger into an
assembly drawing in the space below similar to one of the
examples at the left with instruments or by sketching as
assigned. Space is tight, but you can do it. Use your
creativity; if not now, when?

ASSEMBLY DESIGN

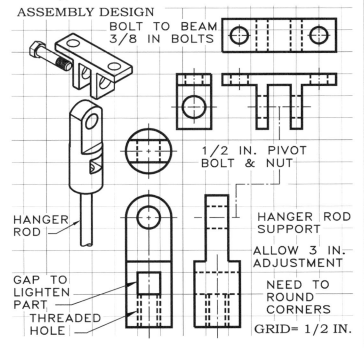

BOLT TO BEAM
3/8 IN BOLTS

1/2 IN. PIVOT
BOLT & NUT

HANGER
ROD

HANGER ROD
SUPPORT

ALLOW 3 IN.
ADJUSTMENT

NEED TO
ROUND
CORNERS

GAP TO
LIGHTEN
PART

THREADED
HOLE

GRID= 1/2 IN.

Graphics & Design
Working Drawings

FILE: . . . . . NAME: . . . .
SEC: . . . . . DATE: . . . . MIN: . . . . .

GRADE

138

Graphics & Design
Assignment Solution

FILE: NAME:
SEC: DATE: MIN:

GRADE

A. Pictorial Assembly: Assembled

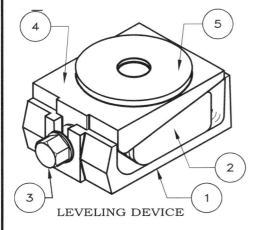

LEVELING DEVICE

B. Exploded Orthographic Assembly

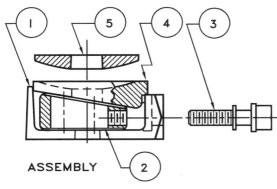

ASSEMBLY

C. Freehand Orthographic Sketch of a Part

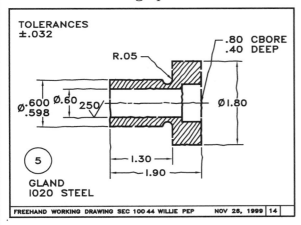

TOLERANCES
±.032

R.05

.80 CBORE
.40 DEEP

Ø.600
.598
Ø.60
250
Ø1.80

1.30
1.90

(5) GLAND
1020 STEEL

FREEHAND WORKING DRAWING SEC 100 44 WILLIE PEP   NOV 28, 1999 14

D. An Exploded Pictorial Assembly

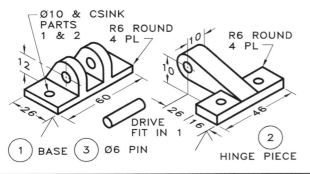

Ø10 & CSINK
PARTS
1 & 2
R6 ROUND
4 PL

12

26

60

DRIVE
FIT IN 1

(1) BASE (3) Ø6 PIN

10
10
26
16
46

R6 ROUND
4 PL

(2)

HINGE PIECE

DESIGN 13: REFINEMENT -- PIPE HANGER
A. Develop the preliminary sketches of the hanger in the space below as freehand refinement drawings that you see as improvements over the given sketches. B. Next, develop your sketches into working drawings with instruments on an A-size sheet. List your design improvements on the sheet in a vacant area.

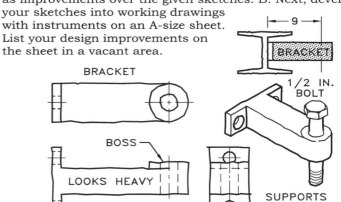

9

BRACKET

BRACKET

1/2 IN.
BOLT

BOSS

LOOKS HEAVY

BOLT TO I-BEAM

SUPPORTS
A HANGING
PIPE

Graphics & Design
Working Drawings: Refinement

FILE: . . . . . NAME: . . . .
SEC: . . . . . DATE: . . . . MIN: . . . . .

GRADE

139

INSTRUCTIONS:
SHEET 1: Draw and dimension parts 1 thru
7 on Size A sheets and dimension each so they
may be fabricated or specified for purchase.
Provide the necessary notes and specifications.

SHEET 2: Draw and dimension parts as an
assembly showing how the parts relate to
each other. This view can be assembled or
exploded to show the relationship of the parts.

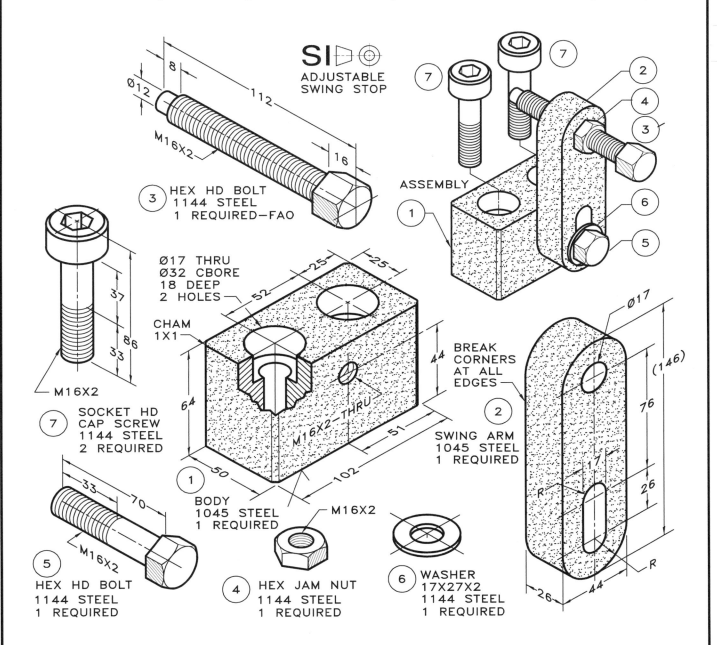

SI◁▷⊙
ADJUSTABLE
SWING STOP

Ø12
8
112
16
M16X2

③ HEX HD BOLT
1144 STEEL
1 REQUIRED—FAO

ASSEMBLY

⑦ ② ④ ③ ⑥ ⑤ ①

Ø17 THRU
Ø32 CBORE
18 DEEP
2 HOLES

25 25
52
CHAM
1X1
44
64
M16X2-THRU
51
50
102

BREAK
CORNERS
AT ALL
EDGES

37
86
33
M16X2

⑦ SOCKET HD
CAP SCREW
1144 STEEL
2 REQUIRED

① BODY
1045 STEEL
1 REQUIRED

② SWING ARM
1045 STEEL
1 REQUIRED

Ø17
(146)
76
17
26
R
R
26 44

33 70
M16X2

⑤ HEX HD BOLT
1144 STEEL
1 REQUIRED

M16X2

④ HEX JAM NUT
1144 STEEL
1 REQUIRED

⑥ WASHER
17X27X2
1144 STEEL
1 REQUIRED

THOUGHT QUESTIONS:

1. What is the purpose of the washer (part 6) in
this assembly?

2. Why is there a slot in the swing arm (part 2)
instead of a circular hole?

3. Why did the designer of this assembly use
both socket-head cap screws and hex-head cap
screws instead of using one or the other?

4. When the swing stop is assembled, what is
the range of adjustment from the upper surface
of the body (part 1) to the centerline of the hex-
head bolt (part 3).

5. What is the maximum distance that the 12
DIA end of the hex-head bolt (part 3) can extend
beyond the face of the swing arm.

6. Write a paragraph giving a description of
either part 1 or part 2 in the absence of a
drawing.

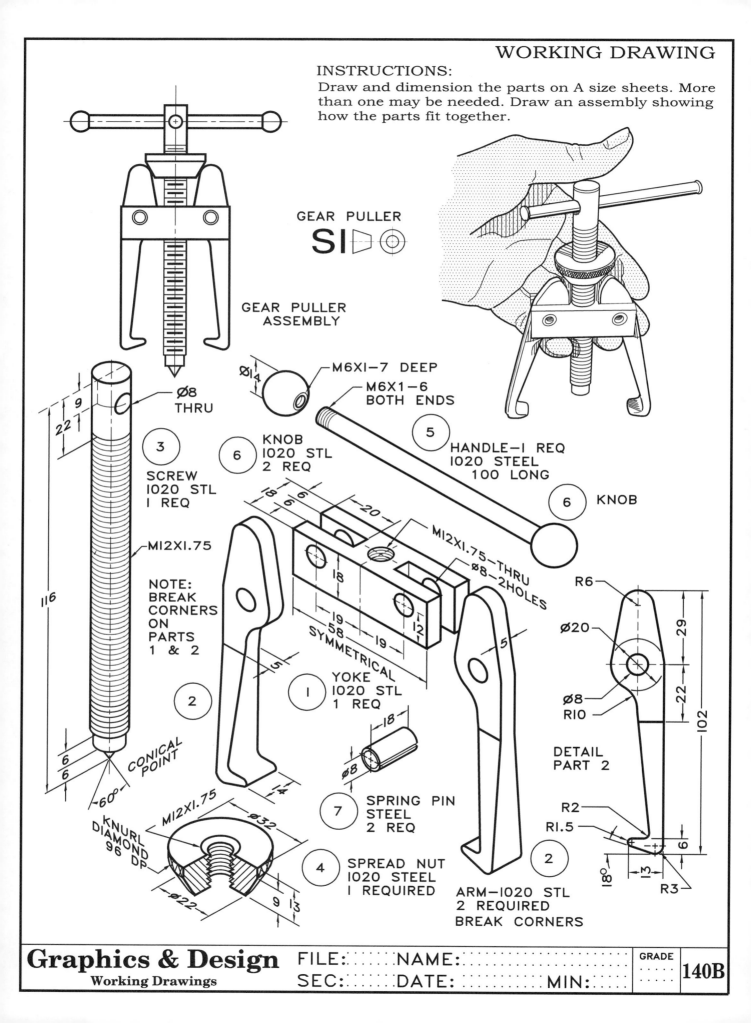

INSTRUCTIONS:
Draw and dimension the parts on A size sheets. More than one may be needed. Draw an assembly showing how the parts fit together.

GEAR PULLER

SI

GEAR PULLER
ASSEMBLY

Ø8
THRU

9
22

3

SCREW
1020 STL
1 REQ

M12X1.75

NOTE:
BREAK
CORNERS
ON
PARTS
1 & 2

116

6
6

CONICAL
POINT

60°

Ø14

M6X1—7 DEEP

M6X1—6
BOTH ENDS

6
KNOB
1020 STL
2 REQ

5

HANDLE—1 REQ
1020 STEEL
100 LONG

6
KNOB

18
6
6
20

M12X1.75—THRU
ø8—2HOLES

18

19
58
19
12

SYMMETRICAL

1
YOKE
1020 STL
1 REQ

2

5

5

R6

Ø20

Ø8
R10

DETAIL
PART 2

29

22

102

R2

R1.5

18°

3

6

R3

2

18

ø8

7
SPRING PIN
STEEL
2 REQ

4

M12X1.75

KNURL
DIAMOND
96 DP

ø32

ø22

9 13

4
SPREAD NUT
1020 STEEL
1 REQUIRED

ARM—1020 STL
2 REQUIRED
BREAK CORNERS

Graphics & Design
Working Drawings

FILE:
SEC:

NAME:
DATE:

GRADE

MIN:

140B

INSTRUCTIONS:
Draw and dimension each part in accordance with the following instructions:
A. SHEET 1: Draw, dimension, and note parts 1 and 3 on a B-zize sheet with an appropriate title block that will be applied to sheets 2 and 3.
B. SHEET 2: Same as sheet 1 showing the remaining parts: 2 plus 4 through 9.
C. SHEET 3: Draw an assembly which shows how the parts relate to each other.

ALTERNATIVE:
Same as above, but draw in A-size sheets instead of B-size sheets.

SLOT 3X5 DEEP

30
23
M10X1.5
2
60°

NOTE:
THE DIMENSION
OF 30 LOCATES THE
THEORETICAL POINT

6 SET SCREW
SLOTTED HEAD
CONICAL POINT
STEEL—2 REQ

WHEN DRAWING,
SHOW FILLETS &
ROUNDS OF R1 ON
ALL CORNERS THAT
DO NOT JOIN FINISHED
SURFACES

Ø20 BASIC
CHAM 1X1
BOTH ENDS
38

Ø26 ±0.2
19

CONICAL HOLES
BOTH SIDES

Ø5
60°

4 SLEEVE—FAO
1020 STEEL
1 REQUIRED

Ø12 ±0.2 THRU
12

Ø20 BASIC

CHAM 1X1
BOTH SIDES

5 BUSHING—FAO
BRASS
2 REQUIRED

M10X1.5 BOTH SIDES
12
66
12
Ø20 BASIC
41
10
20
18
33
Ø38
2
18

2 FORK—1020 STEEL
1 REQUIRED

PART 2 HAS
NO FINISHED
SURFACES

6

7 M10X1.5
REG HEX NUT
STEEL
2 REQUIRED

Ø20 BASIC
CHAMFER
2X2
BOTH
ENDS

9 HEX SOC HEAD
SET SCR
M7X1
9 LONG
STEEL
1 REQ

80

3 POST—1020 STEEL
FAO—1 REQUIRED

M9X1.25

8 HEX SOC HEAD
SET SCR
M9X1.25
9 LONG
STEEL
1 REQ

CYLINDRICAL FITS: H=Hole, d & u=Diameters
PART 1 & PART 3: H9/d9 20.000—20.052/19.883—19.935
PART 2 & PART 3: H9/d9 20.000—20.052/19.883—19.935
PART 4 & PART 5: H7/u6 20.000—20.021/20.042—20.054

116
Ø58
Ø36
Ø20 BASIC
M9X1.25
8
R
6
R12
R24
13
16
37
8
37
16
R
R
R
64

1 BASE—1020 STEEL
1 REQUIRED

SHAFT SUPPORT
SI

INSTRUCTIONS:
Draw and dimension the parts on either A-size sheets or B-size sheets.
Give specifications adequate for the parts to be fabricated. Draw an
asssembly showing how the parts fit together and the necessary title
blocks.

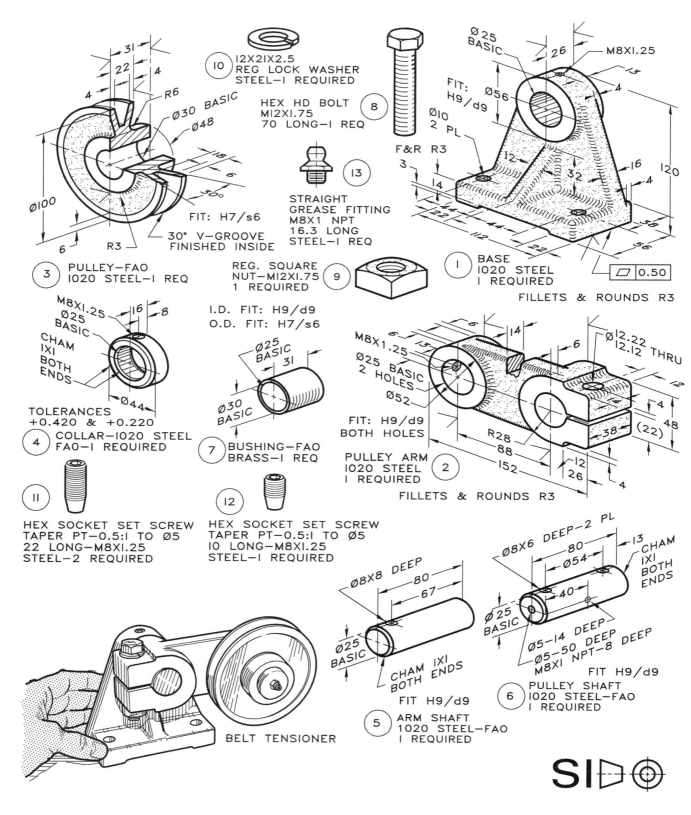

⑩ 12X21X2.5
REG LOCK WASHER
STEEL—1 REQUIRED

HEX HD BOLT
M12X1.75
70 LONG—1 REQ ⑧

⑬ STRAIGHT
GREASE FITTING
M8X1 NPT
16.3 LONG
STEEL—1 REQ

REG. SQUARE
NUT—M12X1.75 ⑨
1 REQUIRED

31
22   4
4   R6
Ø30 BASIC
Ø48
Ø100
FIT: H7/s6
6
30°
18
6
R3
30° V—GROOVE
FINISHED INSIDE

③ PULLEY—FAO
1020 STEEL—1 REQ

Ø25
BASIC   26   M8X1.25
FIT:
H9/d9   Ø56   13
Ø10   4
2 PL
F&R R3
3   12   16   120
14   32   4
22   44   38
44   56
22

① BASE
1020 STEEL
1 REQUIRED

FILLETS & ROUNDS R3

□ 0.50

M8X1.25   16   8
Ø25
BASIC
CHAM
1X1
BOTH
ENDS
Ø44
TOLERANCES
+0.420 & +0.220

④ COLLAR—1020 STEEL
FAO—1 REQUIRED

I.D. FIT: H9/d9
O.D. FIT: H7/s6

Ø25
BASIC   31
Ø30
BASIC

⑦ BUSHING—FAO
BRASS—1 REQ

6   14   6
M8X1.25   Ø12.22 THRU
12.12
Ø25 BASIC
2 HOLES   Ø25 BASIC
Ø52
FIT: H9/d9   4
BOTH HOLES   48
R28   38
88   (22)
PULLEY ARM   12
1020 STEEL   26
1 REQUIRED ②   4
152

FILLETS & ROUNDS R3

⑪
HEX SOCKET SET SCREW
TAPER PT—0.5:1 TO Ø5
22 LONG—M8X1.25
STEEL—2 REQUIRED

⑫
HEX SOCKET SET SCREW
TAPER PT—0.5:1 TO Ø5
10 LONG—M8X1.25
STEEL—1 REQUIRED

Ø8X8 DEEP   80
67
Ø25
BASIC
CHAM 1X1
BOTH ENDS
FIT H9/d9

⑤ ARM SHAFT
1020 STEEL—FAO
1 REQUIRED

Ø8X6 DEEP—2 PL   80   13
Ø54   CHAM
40   1X1
BOTH
Ø25   ENDS
BASIC
Ø5—14 DEEP
Ø5—50 DEEP
M8X1 NPT—8 DEEP
FIT H9/d9

⑥ PULLEY SHAFT
1020 STEEL—FAO
1 REQUIRED

BELT TENSIONER

SI◁⊙

Graphics & Design
Standard Keys and Keyways

FILE: ⋯⋯ NAME: ⋯⋯⋯⋯
SEC: ⋯⋯ DATE: ⋯⋯ MIN: ⋯⋯

GRADE

141B

INSTRUCTIONS:
Draw and dimension each part in accordance with the following instructions:
A. SHEET 1: Draw, dimension, and note parts 1 and 3 on a B-size sheet with an appropriate title block that will be applied to sheets 2 and 3 also.
B. SHEET 2: Same as sheet 1 showing the remaining parts: 2 plus 4 through 6.
C. SHEET 3: Draw an assembly which shows how the parts relate to each other. Draw the parts list on sheet 3.

CYLINDRICAL FITS: H=Hole,   H=Hole, c & s=Shafts
PART 4 & PART 3: H11/c11 32.120−32.280/31.720−31.880
PART 3 & PART 5: H7/s6   40.000−40.025/40.043−40.059
PART 2 & PART 4: H7/s6   24.000−24.021/24.035−24.048

ROLLER SUPPORT ASSEMBLY

GREASE FITTING
M6X1−25 LONG
STEEL−1 REQ

M18X2.5−60 LONG
HEX HD SCREW
4 REQUIRED

Ø40
Ø32
BUSHING
BRASS
2 REQUIRED
FAO

92
32 30

Ø160
Ø70 Ø48
Ø48
8

BELL ROLLER
1010 STEEL
1 REQ

Ø40 THRU

FILLETS &
ROUNDS R4

6
Ø48
BUSHING
BRASS
2 REQUIRED
FAO

Ø24
Ø48
Ø24
TYP
120
Ø20−2
HOLES
6
16

BRACKET
CAST IRON
2 REQ

Ø20−4 HOLES
72
108
20

28
56
20
R TYP
8
R6

260 224
108
6
108
6

R−4 PL

BASE PLATE
1010 STEEL
1 REQUIRED

36
108 36
76

BRACKET−CAST IRON
2 REQUIRED

ROLLER SUPPORT
SI

2X2 CHAMFER
28 54   160 54
Ø3
THRU
Ø32
Ø24
Ø5−90 DEEP
M6X1−12 DEEP

SHAFT−1010 STEEL
1 REQUIRED−FAO

Graphics & Design
Working Drawing

FILE: . . . . NAME: . . . .
SEC: . . . . . . DATE: . . . . MIN: . . . .

GRADE

142

INSTRUCTIONS:
Use these partial views and key dimensions to make detail drawings
of the parts of this assembly as if you were its designer. Provide the
essential details, specifications, and dimensions of each part plus an
assembly drawing. Draw on A-size or B-size sheets.

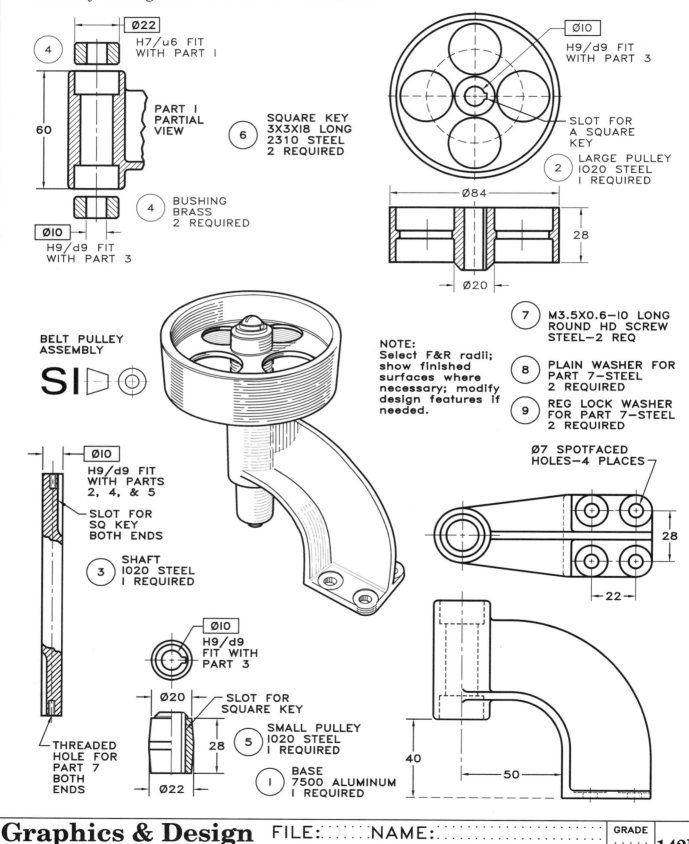

④ Ø22
H7/u6 FIT
WITH PART 1

60

PART 1
PARTIAL
VIEW

⑥ SQUARE KEY
3X3X18 LONG
2310 STEEL
2 REQUIRED

④ BUSHING
BRASS
2 REQUIRED

Ø10
H9/d9 FIT
WITH PART 3

Ø10
H9/d9 FIT
WITH PART 3

SLOT FOR
A SQUARE
KEY

② LARGE PULLEY
1020 STEEL
1 REQUIRED

Ø84

28

Ø20

BELT PULLEY
ASSEMBLY

SI

NOTE:
Select F&R radii;
show finished
surfaces where
necessary; modify
design features if
needed.

⑦ M3.5X0.6-10 LONG
ROUND HD SCREW
STEEL-2 REQ

⑧ PLAIN WASHER FOR
PART 7-STEEL
2 REQUIRED

⑨ REG LOCK WASHER
FOR PART 7-STEEL
2 REQUIRED

Ø10
H9/d9 FIT
WITH PARTS
2, 4, & 5

SLOT FOR
SQ KEY
BOTH ENDS

③ SHAFT
1020 STEEL
1 REQUIRED

Ø7 SPOTFACED
HOLES-4 PLACES

28

22

Ø10
H9/d9
FIT WITH
PART 3

THREADED
HOLE FOR
PART 7
BOTH
ENDS

Ø20 SLOT FOR
SQUARE KEY

⑤ SMALL PULLEY
1020 STEEL
1 REQUIRED

28

Ø22

① BASE
7500 ALUMINUM
1 REQUIRED

40

50

Graphics & Design
Working Drawing: Design

FILE: NAME:
SEC: DATE: MIN:

GRADE

142B

A. An Oblique Drawing

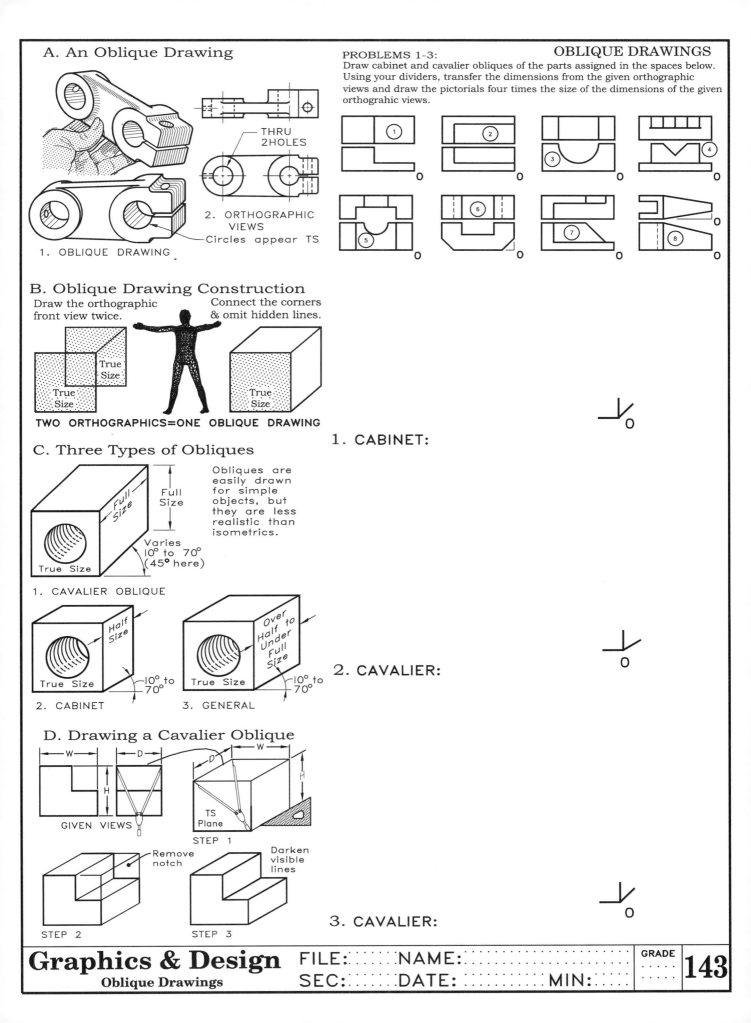

1. OBLIQUE DRAWING

2. ORTHOGRAPHIC VIEWS
— THRU 2HOLES
— Circles appear TS

Draw cabinet and cavalier obliques of the parts assigned in the spaces below. Using your dividers, transfer the dimensions from the given orthographic views and draw the pictorials four times the size of the dimensions of the given orthograhic views.

B. Oblique Drawing Construction

Draw the orthographic front view twice.

Connect the corners & omit hidden lines.

True Size

True Size

True Size

TWO ORTHOGRAPHICS=ONE OBLIQUE DRAWING

C. Three Types of Obliques

Full Size

Full Size

True Size

Varies 10° to 70° (45° here)

1. CAVALIER OBLIQUE

Obliques are easily drawn for simple objects, but they are less realistic than isometrics.

Half Size

True Size

10° to 70°

2. CABINET

Over Half to Under Full Size

True Size

10° to 70°

3. GENERAL

D. Drawing a Cavalier Oblique

W   D

H

GIVEN VIEWS

W

H

TS Plane

STEP 1

Remove notch

STEP 2

Darken visible lines

STEP 3

1. CABINET:

2. CAVALIER:

3. CAVALIER:

Graphics & Design
Oblique Drawings

FILE: .... NAME: ....
SEC: .... DATE: .... MIN: ......

GRADE

143

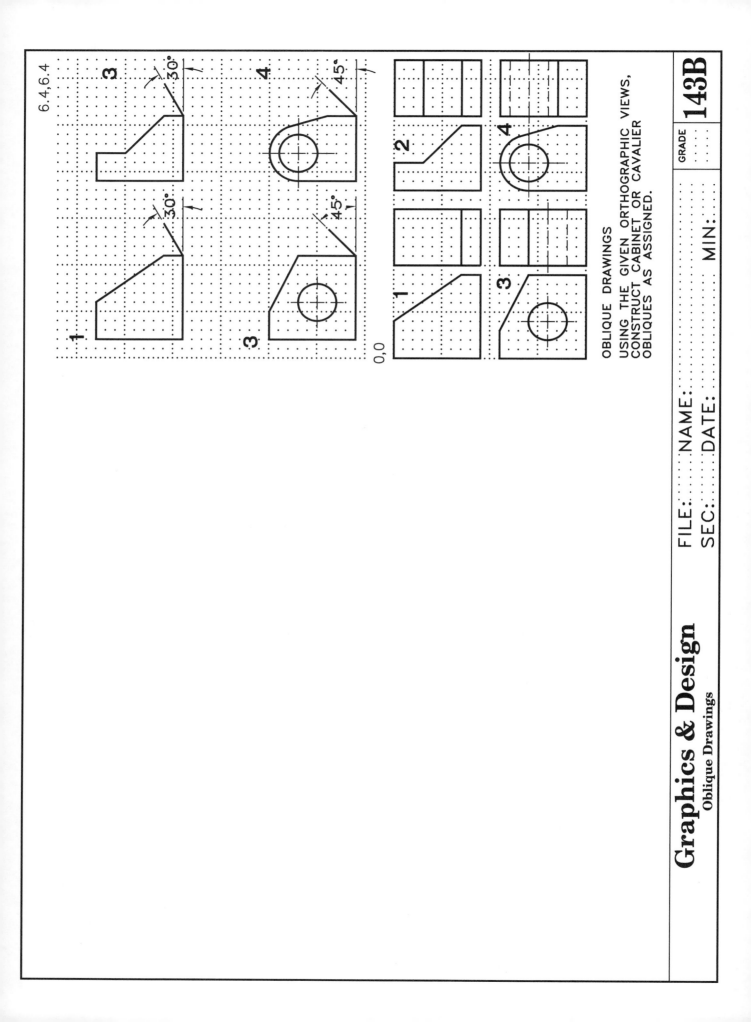

6.4,6.4

3

30°

1

30°

4

45°

3

45°

2

0,0

1

4

3

OBLIQUE DRAWINGS

USING THE GIVEN ORTHOGRAPHIC VIEWS,
CONSTRUCT CABINET OR CAVALIER
OBLIQUES AS ASSIGNED.

Graphics & Design
Oblique Drawings

FILE: NAME:

SEC: DATE:

GRADE 143B MIN:

A. Isometric Drawing

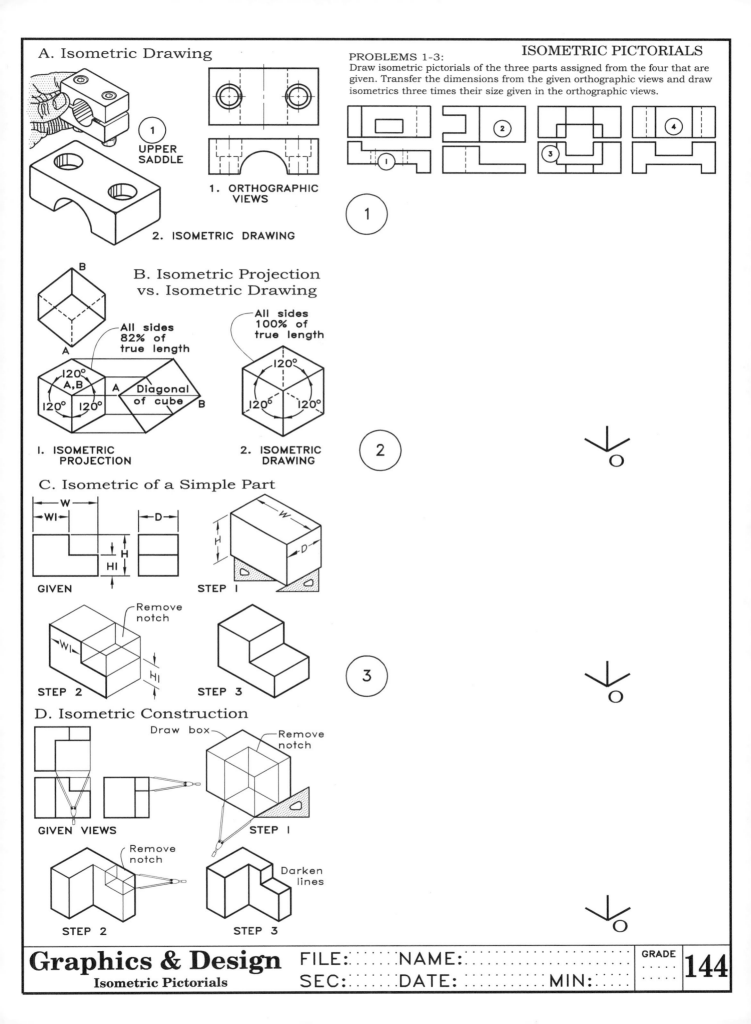

UPPER
SADDLE

1. ORTHOGRAPHIC
VIEWS

2. ISOMETRIC DRAWING

PROBLEMS 1-3:
Draw isometric pictorials of the three parts assigned from the four that are given. Transfer the dimensions from the given orthographic views and draw isometrics three times their size given in the orthographic views.

ISOMETRIC PICTORIALS

B. Isometric Projection vs. Isometric Drawing

All sides 82% of true length

120°
A,B
120° 120°
A
Diagonal of cube
B

I. ISOMETRIC PROJECTION

All sides 100% of true length

120°
120° 120°

2. ISOMETRIC DRAWING

C. Isometric of a Simple Part

GIVEN

STEP I

Remove notch

STEP 2

STEP 3

D. Isometric Construction

GIVEN VIEWS

Draw box

Remove notch

STEP I

Remove notch

STEP 2

Darken lines

STEP 3

Graphics & Design
Isometric Pictorials

FILE: NAME:
SEC: DATE: MIN:

GRADE

144

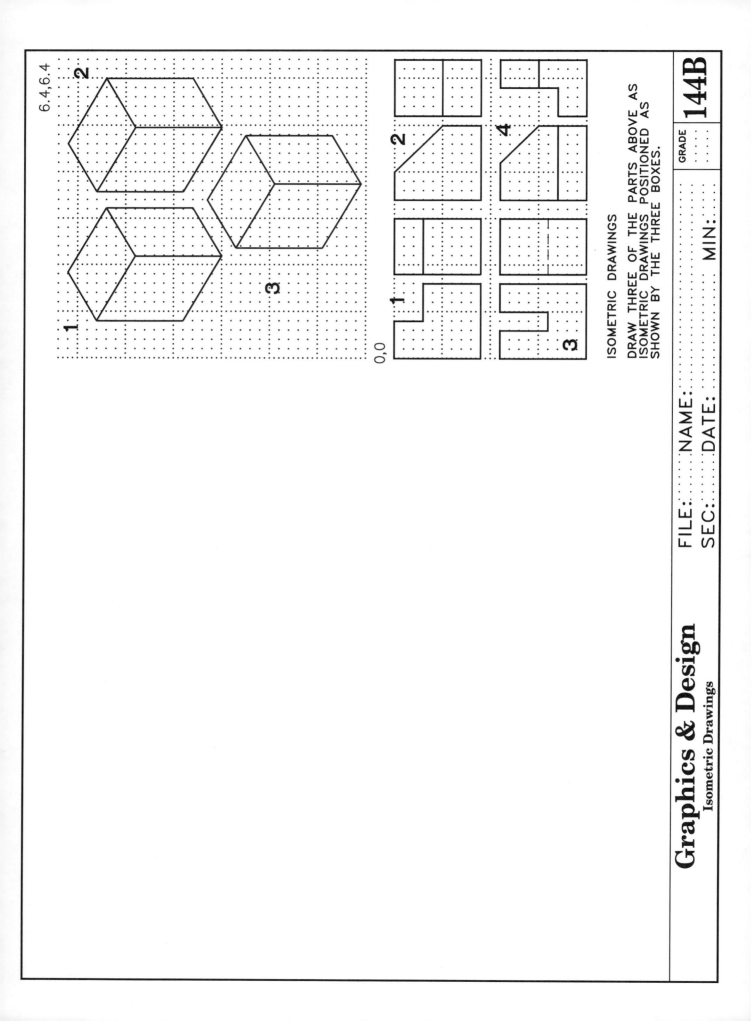

6.4,6.4

ISOMETRIC DRAWINGS

DRAW THREE OF THE PARTS ABOVE AS
ISOMETRIC DRAWINGS POSITIONED AS
SHOWN BY THE THREE BOXES.

Graphics & Design
Isometric Drawings

FILE: NAME:
SEC: DATE: MIN:

GRADE **144B**

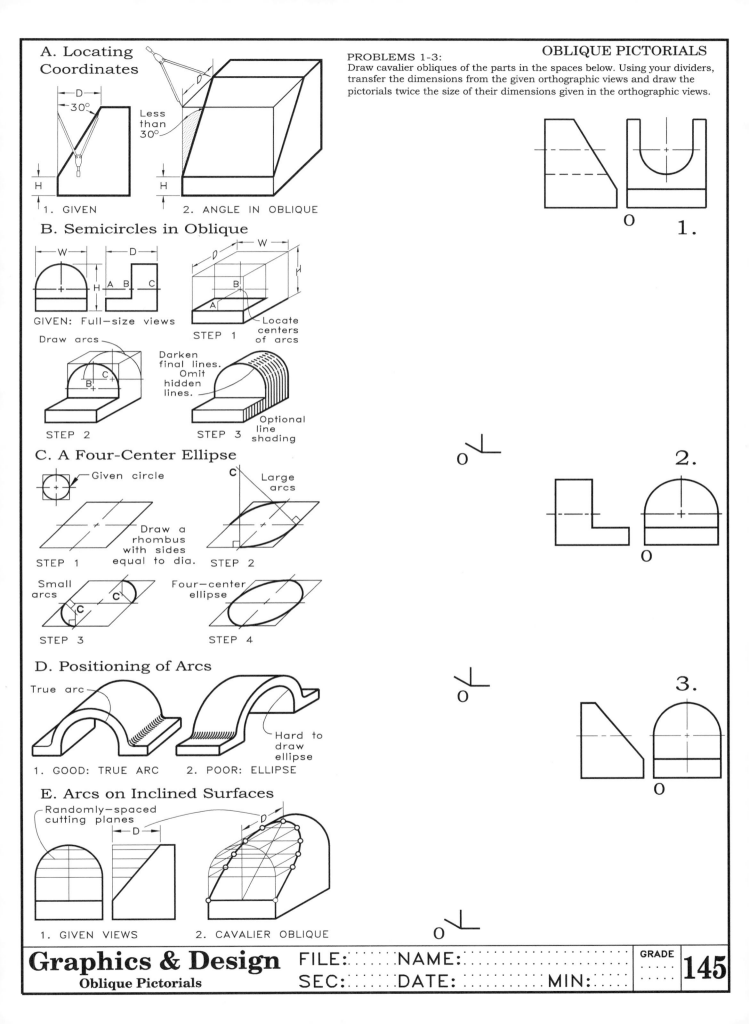

A. Locating Coordinates

1. GIVEN

2. ANGLE IN OBLIQUE

B. Semicircles in Oblique

GIVEN: Full-size views

STEP 1 — Locate centers of arcs

STEP 2 — Draw arcs

STEP 3 — Darken final lines. Omit hidden lines. Optional line shading

C. A Four-Center Ellipse

Given circle

STEP 1 — Draw a rhombus with sides equal to dia.

STEP 2 — Large arcs

STEP 3 — Small arcs

STEP 4 — Four-center ellipse

D. Positioning of Arcs

True arc

1. GOOD: TRUE ARC

2. POOR: ELLIPSE — Hard to draw ellipse

E. Arcs on Inclined Surfaces

Randomly-spaced cutting planes

1. GIVEN VIEWS

2. CAVALIER OBLIQUE

Draw cavalier obliques of the parts in the spaces below. Using your dividers, transfer the dimensions from the given orthographic views and draw the pictorials twice the size of their dimensions given in the orthographic views.

0   1.

0   2.

0   3.

Graphics & Design
Oblique Pictorials

A. Inclined Surfaces

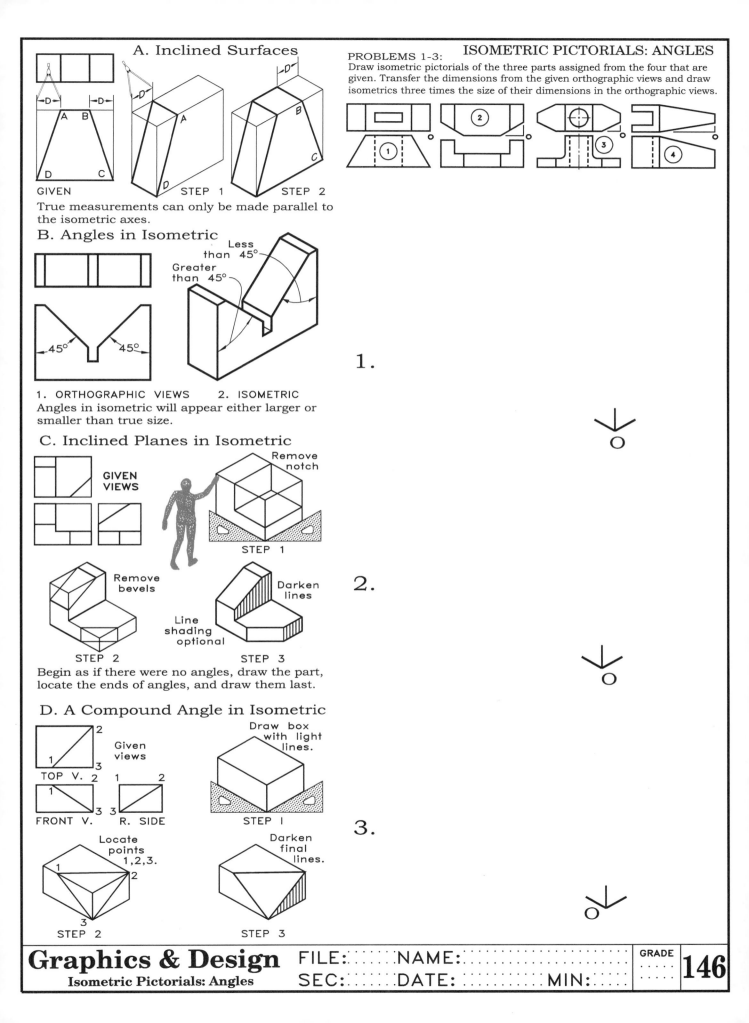

GIVEN   STEP 1   STEP 2

True measurements can only be made parallel to the isometric axes.

B. Angles in Isometric

Less than 45°

Greater than 45°

45°   45°

1. ORTHOGRAPHIC VIEWS   2. ISOMETRIC

Angles in isometric will appear either larger or smaller than true size.

C. Inclined Planes in Isometric

GIVEN VIEWS

Remove notch

STEP 1

Remove bevels

STEP 2

Darken lines

Line shading optional

STEP 3

Begin as if there were no angles, draw the part, locate the ends of angles, and draw them last.

D. A Compound Angle in Isometric

Given views

TOP V.

FRONT V.   R. SIDE

Draw box with light lines.

STEP 1

Locate points 1,2,3.

STEP 2

Darken final lines.

STEP 3

PROBLEMS 1-3:
Draw isometric pictorials of the three parts assigned from the four that are given. Transfer the dimensions from the given orthographic views and draw isometrics three times the size of their dimensions in the orthographic views.

1   2   3   4

1.

O

2.

O

3.

O

Graphics & Design
Isometric Pictorials: Angles

FILE: NAME:
SEC: DATE: MIN:

GRADE **146**

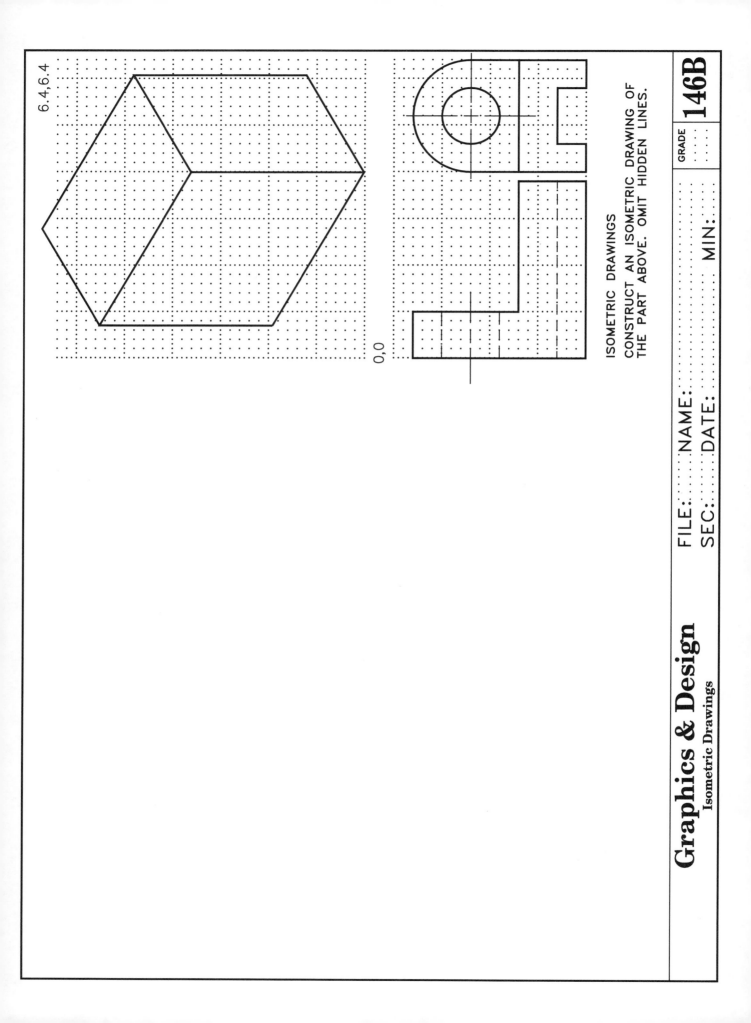

6.4,6.4

0,0

ISOMETRIC DRAWINGS

CONSTRUCT AN ISOMETRIC DRAWING OF
THE PART ABOVE. OMIT HIDDEN LINES.

Graphics & Design
Isometric Drawings

FILE: . . . . . . NAME: . . . .

SEC: . . . . . . DATE: . . . .

A. The Four-Center Ellipse

GIVEN

DIA

Draw a rhombus

DIA

STEP 1

Locate centers

STEP 2

Draw 4 arcs

STEP 3

1.

B. Ellipses on Isometric Surfaces

Four-center ellipses can be drawn on all three isometric planes.

2.

C. Four-Center Cylinder Construction

Isometric axes

Axis

Omit hidden lines

STEP 1   STEP 2   STEP 3

PROBLEMS 1-3:
Draw isometric ellipses within the construction lines given and show your additional construction lines.

D. Ellipse Terminology

Isometric template 2 inch ellipse

Isometric Diameter

1.

True Diameter

Major   Dia
Dia
Min

2.

3.

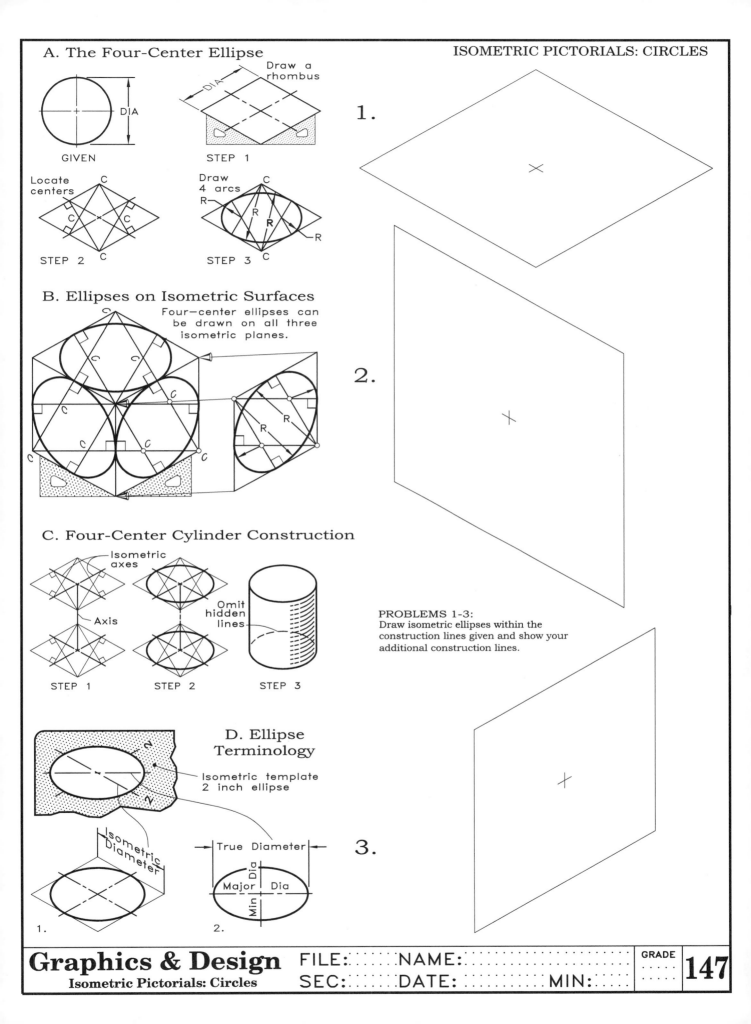

Graphics & Design
Isometric Pictorials: Circles

FILE: NAME:
SEC: DATE: MIN:

GRADE

147

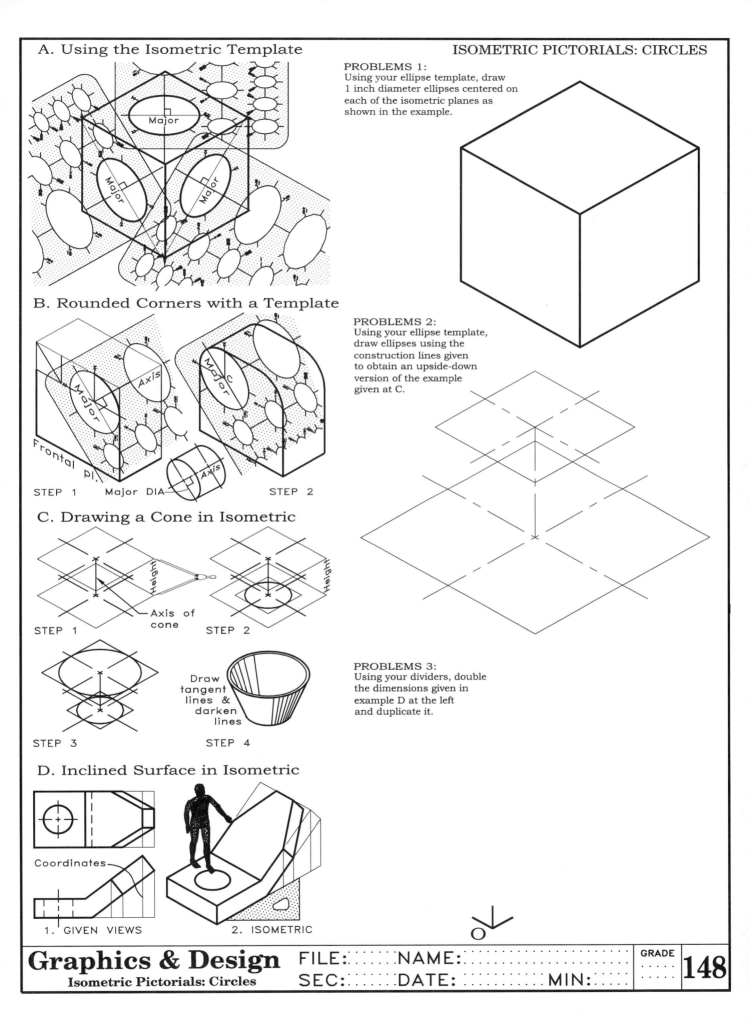

A. Using the Isometric Template

B. Rounded Corners with a Template

STEP 1   Major DIA   STEP 2

C. Drawing a Cone in Isometric

STEP 1   Axis of cone   STEP 2

STEP 3   Draw tangent lines & darken lines   STEP 4

D. Inclined Surface in Isometric

Coordinates

1. GIVEN VIEWS   2. ISOMETRIC

PROBLEMS 1:
Using your ellipse template, draw
1 inch diameter ellipses centered on
each of the isometric planes as
shown in the example.

PROBLEMS 2:
Using your ellipse template,
draw ellipses using the
construction lines given
to obtain an upside-down
version of the example
given at C.

PROBLEMS 3:
Using your dividers, double
the dimensions given in
example D at the left
and duplicate it.

Graphics & Design
Isometric Pictorials: Circles

FILE: NAME:
SEC: DATE: MIN:

GRADE 148

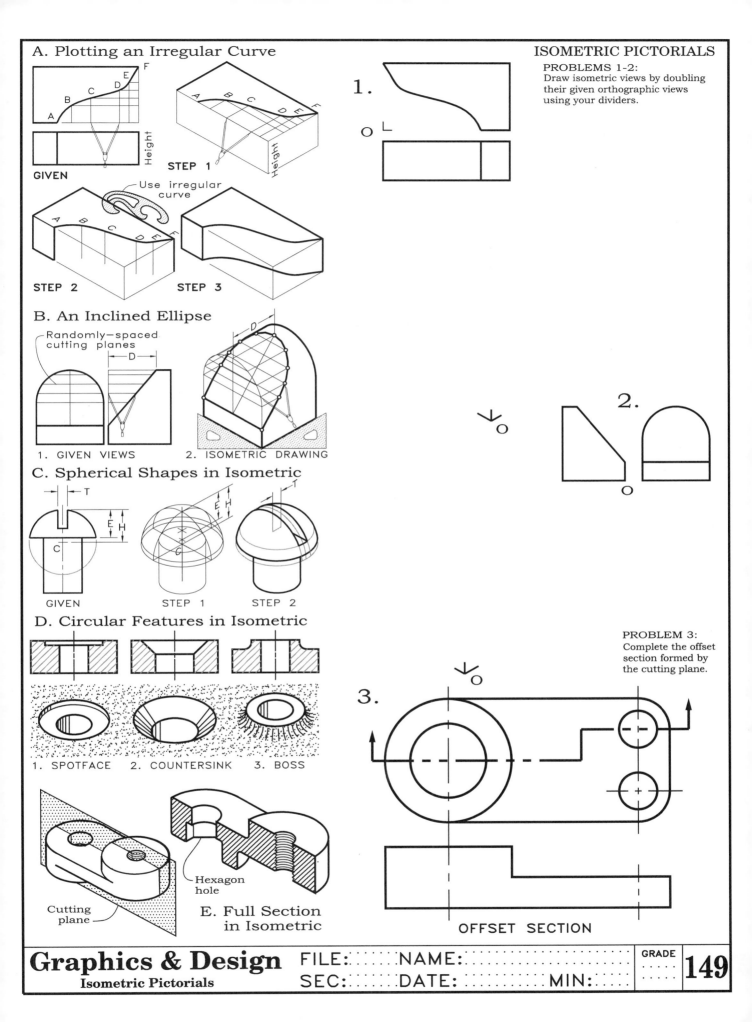

A. Plotting an Irregular Curve

GIVEN

STEP 1

Use irregular curve

STEP 2

STEP 3

B. An Inclined Ellipse

Randomly-spaced cutting planes

1. GIVEN VIEWS

2. ISOMETRIC DRAWING

C. Spherical Shapes in Isometric

GIVEN

STEP 1

STEP 2

D. Circular Features in Isometric

1. SPOTFACE

2. COUNTERSINK

3. BOSS

Cutting plane

Hexagon hole

E. Full Section in Isometric

ISOMETRIC PICTORIALS

PROBLEMS 1-2:
Draw isometric views by doubling their given orthographic views using your dividers.

1.

O

2.

O

PROBLEM 3:
Complete the offset section formed by the cutting plane.

3.

OFFSET SECTION

Graphics & Design
Isometric Pictorials

FILE: .... NAME: ....
SEC: .... DATE: .... MIN: ....

GRADE

149

PRODUCT DEVELOPMENT
Cost Per Unit

LABOR	$40	40%X360°=144°
RESEARCH	30	30%X360°=108°
MATERIALS	20	20%X360°= 72°
OVERHEAD	10	10%X360°= 36°
TOTAL	$100	360°

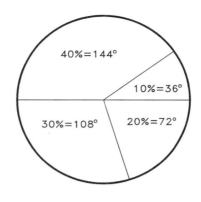

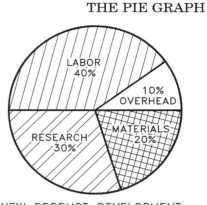

NEW PRODUCT DEVELOPMENT
COST PER UNIT

STEP 1: Find the sum of the parts and the percentage of each. Multiply each percentage by 360° to obtain the angle of each sector.

STEP 2: Draw the circle and construct each sector with the degrees from Step 1. Place slim sectors as nearly horizontal as possible.

STEP 3: Label sectors with their proper names and percentages. Exact numbers may also be included in each sector for clarity.

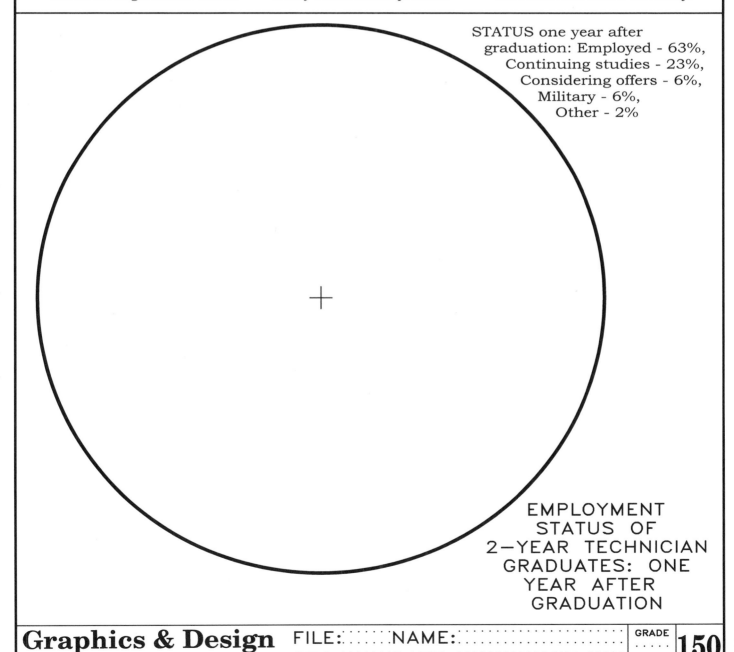

STATUS one year after graduation: Employed - 63%, Continuing studies - 23%, Considering offers - 6%, Military - 6%, Other - 2%

EMPLOYMENT STATUS OF 2−YEAR TECHNICIAN GRADUATES: ONE YEAR AFTER GRADUATION

6.4,6.4

MARKET SHARE OF CAR SALES.

0,0

THE PIE GRAPH

CONSTRUCT A PIE GRAPH THAT SHOWS
THE MARKET SHARE OF THE FOLLOWING
MAKERS:

		DEG.
GENERAL MOTORS	34.3%	123
FORD	16.7%	60
CHRYSLER	13.5%	49
TRANSPLANTS	9.4%	34
IMPORTS	26.1%	94

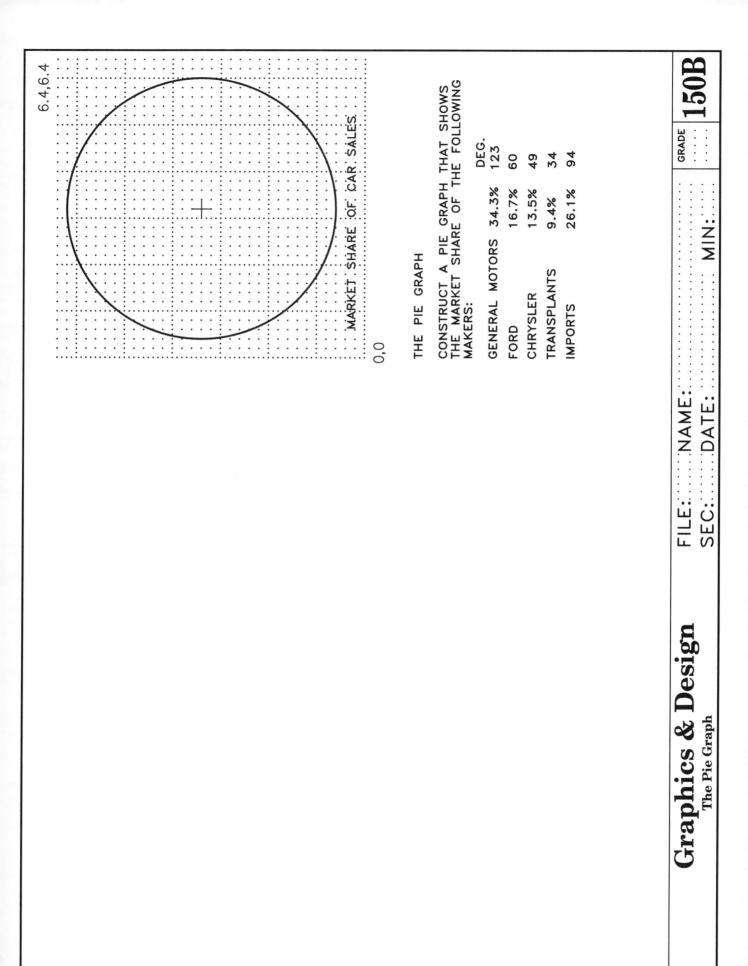

Graphics & Design
The Pie Graph

FILE: ......NAME: ....

SEC: ......DATE: ....

150B

BAR GRAPH

Example

DIVIDENDS PAID BY
THE APEX COMPANY

YEAR	DIVIDEND
2005	$0.40
2006	0.60
2007	0.90
2008	0.80

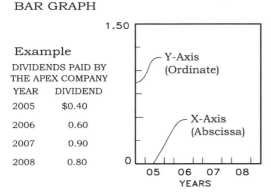

STEP 1: Scale the vertical and horizontal axes so the data will fit on the grid. Begin the bars at zero

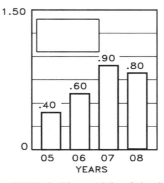

STEP 2: The width of the bars should be greater than the space between them. Lines should not cross the bars.

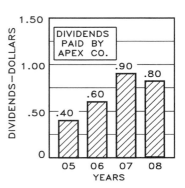

STEP 3: Strengthen all lines, place a title in the graph, label the axes, and crosshatch the bars.

REQUIRED: Plot the data as a bar graph, ranking the numbers of employees of the company by job description starting with the smallest category at the top and ending with the largest at the bottom. Give a title, label and hatch the bars, and strengthen all lines as necessary to finish the bar graph.

WILSON-REYNOLDS MFGR.

Engineers	5%
Research	10%
Mfgr. Staff	62%
Sales Staff	23%

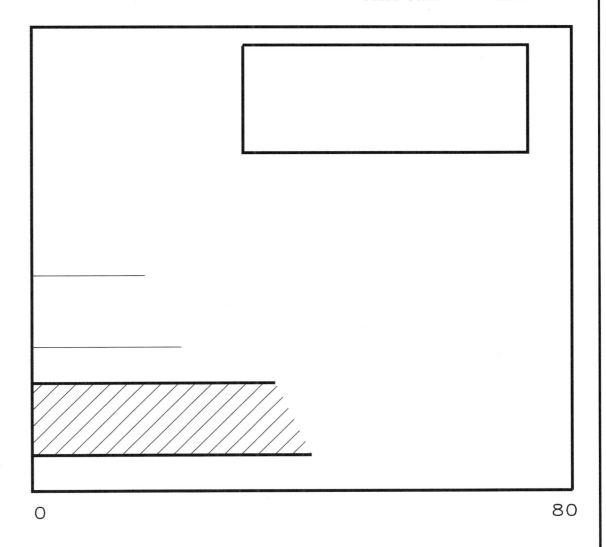

0   80

6.4,6.4

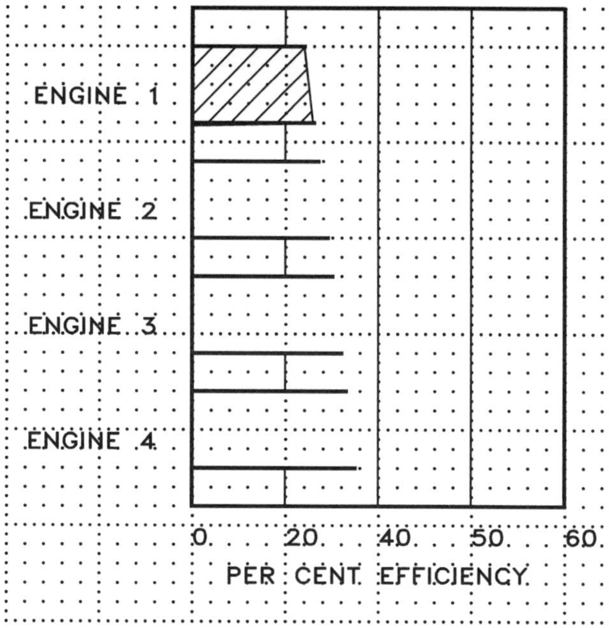

ENGINE 1

ENGINE 2

ENGINE 3

ENGINE 4

0,0

0   10   20   30   40   50   60

PER CENT EFFICIENCY

THE BAR GRAPH

PLOT THE DATA BELOW AS A BAR
GRAPH AND RANK THE ENGINES BY
THEIR EFFICIENCY STARTING WITH THE
THE LEAST EFFICIENT ONE AT THE TOP.
GIVE A TITLE AND LABEL THE AXES.

ENGINE 1   28%
ENGINE 2   54%
ENGINE 3   38%
ENGINE 4   46%

Graphics & Design
The Bar Graph

FILE: NAME:
SEC: DATE:

SMITH INC.

Year	$ Millions	
	Exp	Gross
01	2	1
02	3	2
03	3	2
04	5	3
05	7	6
06	9	11
07	10	13
08	12	15

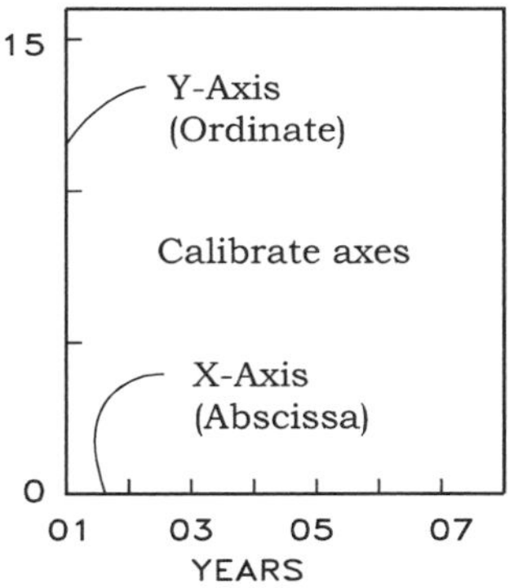

Y-Axis (Ordinate)

Calibrate axes

X-Axis (Abscissa)

STEP 1: Using the given data, draw the vertical (ordinate) and horizontal (abscissa) axes to accomodate the largest values.

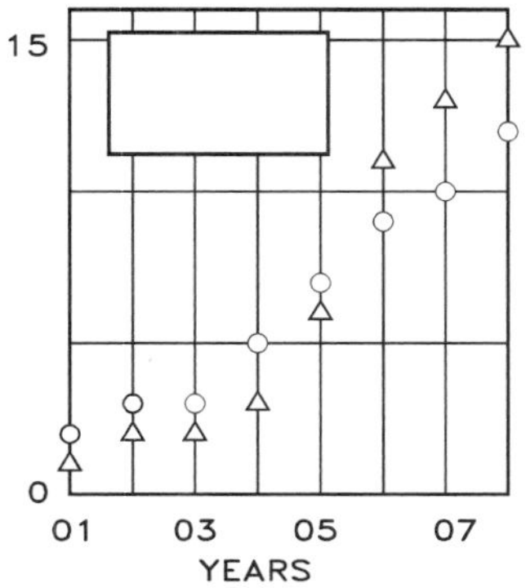

STEP 2: Draw division lines, a title box, and plot the data using different symbols for each set of data.

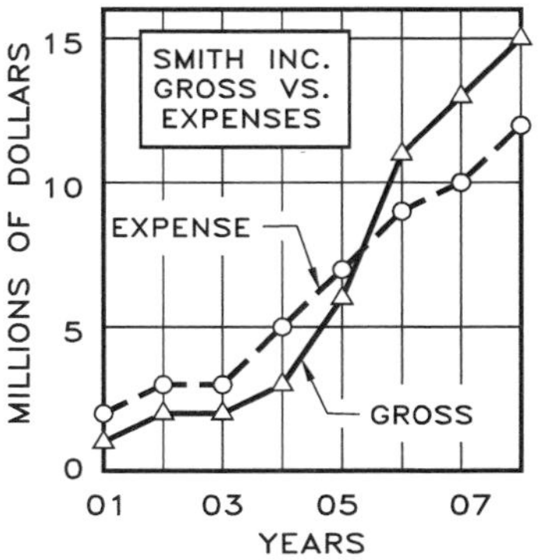

SMITH INC. GROSS VS. EXPENSES

EXPENSE

GROSS

STEP 3: Connect points with straight lines, label the axes, title the graph, darken the lines, and label the curves.

REQUIRED: Using the data from the Ajax Chain, plot its data in the manner shown in the example above. Provide the missing lines and labels needed to complete the graph.

	JAN	FEB	MAR	APR	MAY	JUN	JUL
GROSS	130	170	160	210	320	380	370
EXPENSES	80	130	120	125	230	228	200
MONTH	JAN	FEB	MAR	APR	MAY	JUN	JUL

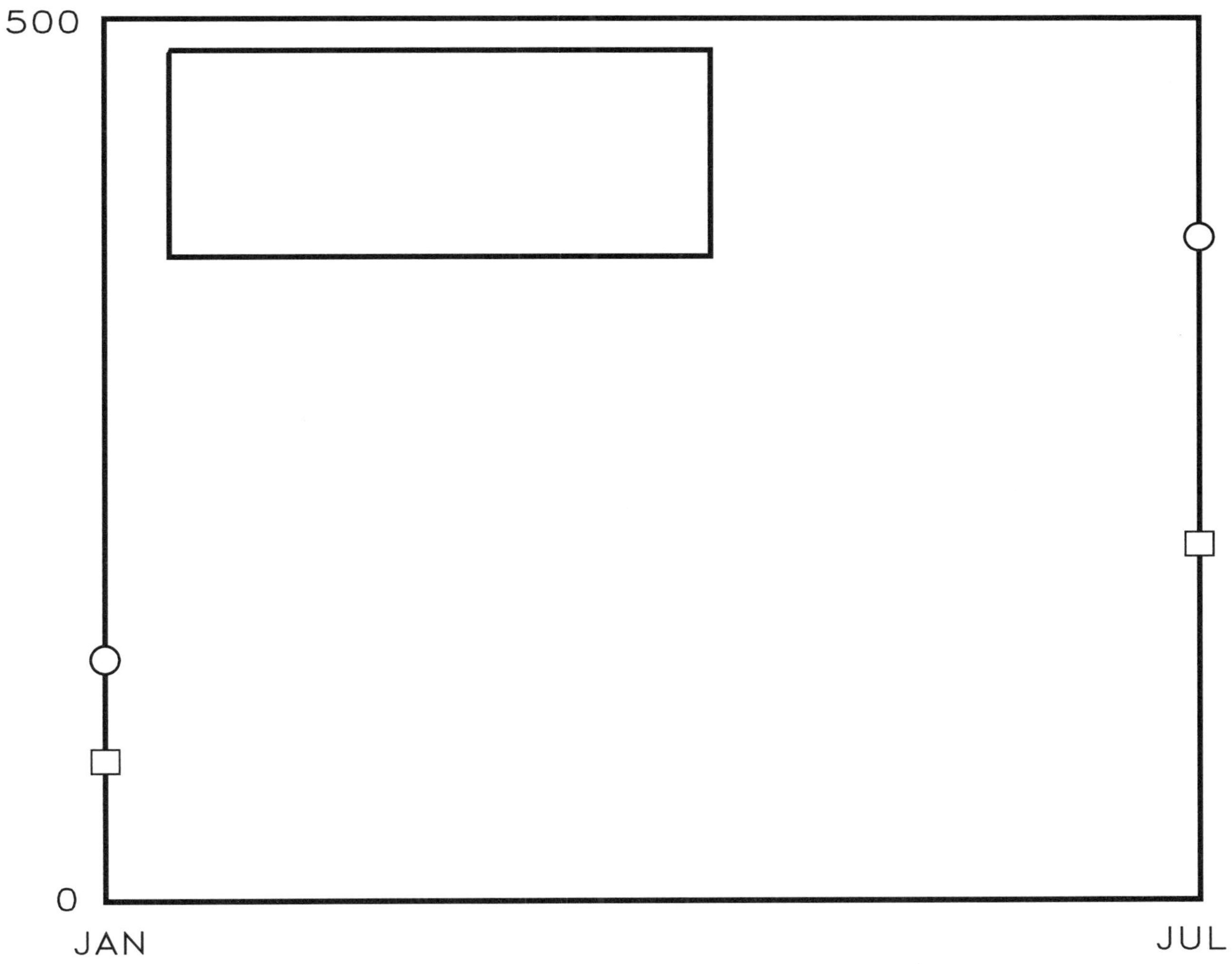

500

0

JAN

JUL

6.4,6.4

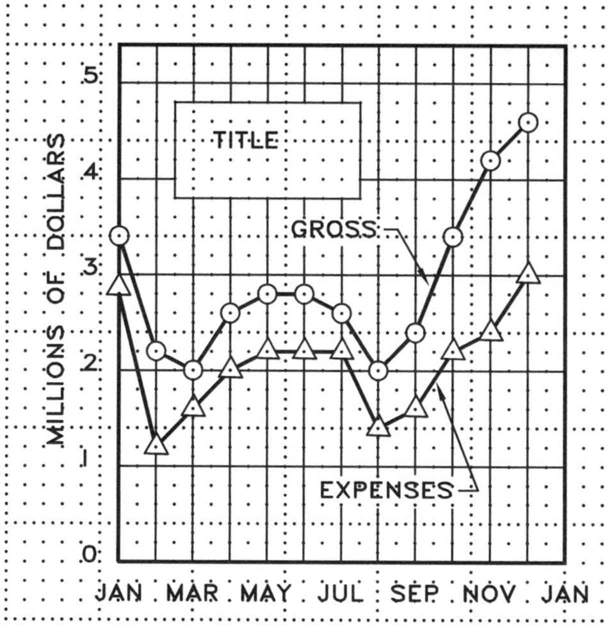

TITLE

GROSS

EXPENSES

.5
.4
.3
.2
.1
.0

MILLIONS OF DOLLARS

JAN MAR MAY JUL SEP NOV JAN

0,0

BROKEN—LINE GRAPH

THE DATA ABOVE SHOWS A COMPARISON
BETWEEN THE GROSS INCOME AND THE
EXPENSES OF OPERATION FOR A TWELVE—
MONTH PERIOD FOR THE APEX SALES
COMPANY.
   DRAW THIS GRAPH USING GOOD
PRACTICES OF GRAPHING. LABEL THE
CURVES, PROVIDE A TITLE, AND LABEL
THE AXES.

Graphics & Design
Broken-Line Graph

FILE: .... NAME: ....
SEC: .... DATE: ....

GRADE .....

152B

MIN: .....

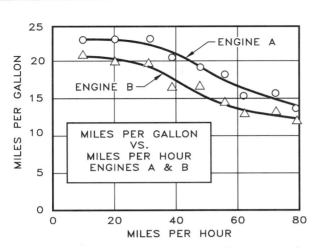

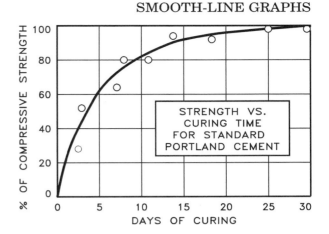

Data can be represented by best-fit curves that approximate the data without passing through each point, since it is known that the data being graphed will yield a smooth-line curve.

When the data being graphed involves gradual and continuous changes in relationships, the curve is drawn as a smooth line and does not necessarily pass through each data point.

REQUIRED: Construct a linear coordinate graph that shows the relationship of energy costs (mills per kilowatt-hour) on the y-axis to the percent capacity of a nuclear plant and a gas-fired plant on the x-axis.

GAS-FIRED PLANT: 17 mills - 10%; 12 mills - 20%; 8 mills - 40%; 7 mills - 60%; 6 mills - 80%; 5.8 mills - 100%. NUCLEAR PLANT: 24 mills - 10%; 14 mills - 20%; 7 mills - 40%; 5 mills - 60%; 4.2 mills - 80%; 3.7 mills - 100%.
(1 mill = 1/10 of a cent)

6.4,6.4

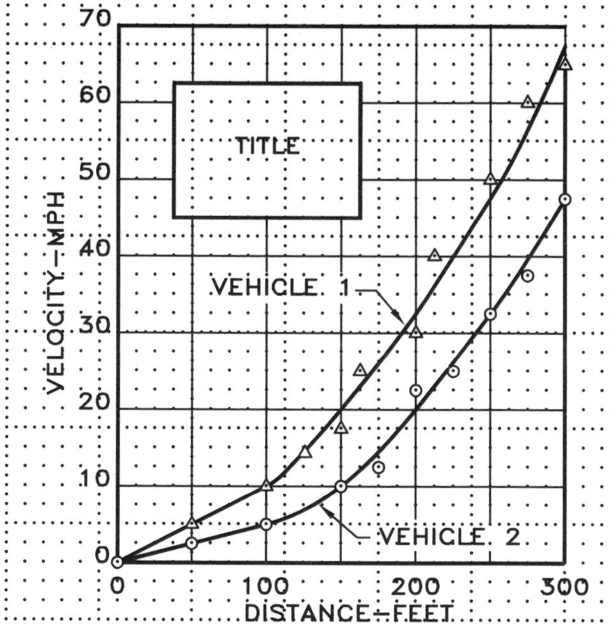

VELOCITY—MPH

70

60

50

40

30

20

10

0

0,0

TITLE

VEHICLE 1

VEHICLE 2

100   200   300
DISTANCE—FEET

SMOOTH—LINE GRAPHS

THE GRAPH ABOVE SHOWS A COMPARISON
BETWEEN THE VELOCITIES OF TWO TEST
VEHICLES AT VARIOUS DISTANCES FROM
THE STARTING POINT.
    DRAW THE GRAPH USING GOOD GRAPH-
ING PRINCIPLES, LABEL THE AXES, AND
PROVIDE A TITLE.

Graphics & Design
Smooth-Line Graphs

FILE: . . . . NAME: . . . .
SEC: . . . . DATE: . . . .

GRADE | 153B
MIN: . . . .

Break-Even Graph

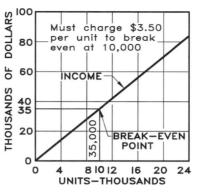

STEP 1: Plot the development cost ($20,000). At $1.50 per unit to make, the total cost would be $35,000 for 10,000 units, the break-even point.

STEP 2: To break even at 10,000, the manufacturer must sell each for $3.50. Draw a line from zero through the break-even point ($35,000) to represent income.

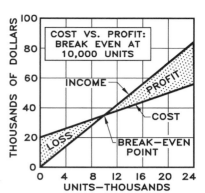

STEP 1: There is a loss of $20,000 at zero, but progressively less until reaching the break-even point. Profit is the difference between the two lines at the right of the of the break-even point.

REQUIRED: Our company wishes to detemine the financial prospects of the belt tensioner, and needs a break-even graph for analysis purposes. We must spend $24,000 on development and each tensioner is expected to cost $40 to produce. Draw a break-even graph to price the product if it is to break even when 400 units are sold. Answer the questions at the right.

What will be:
The retail price?
Profit at 700 units?
Loss at 300 units?

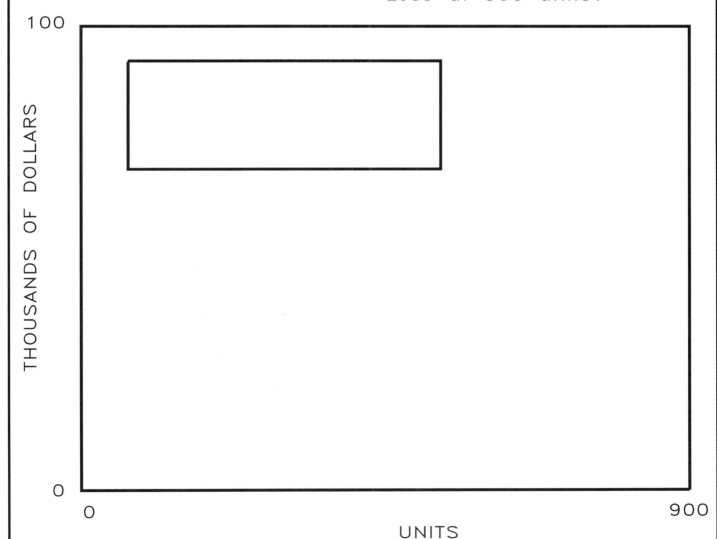

THOUSANDS OF DOLLARS

UNITS

BREAK-EVEN GRAPH

BY FOLLOWING THE INSTRUCTIONS ON THE
OPPOSITE SIDE OF THE PAGE, DRAW A
BREAK-EVEN GRAPH. COMPLETE THE TABLE
OF VALUES AND GIVE THE ANSWERS.

Graphics & Design
Break-Even Graph

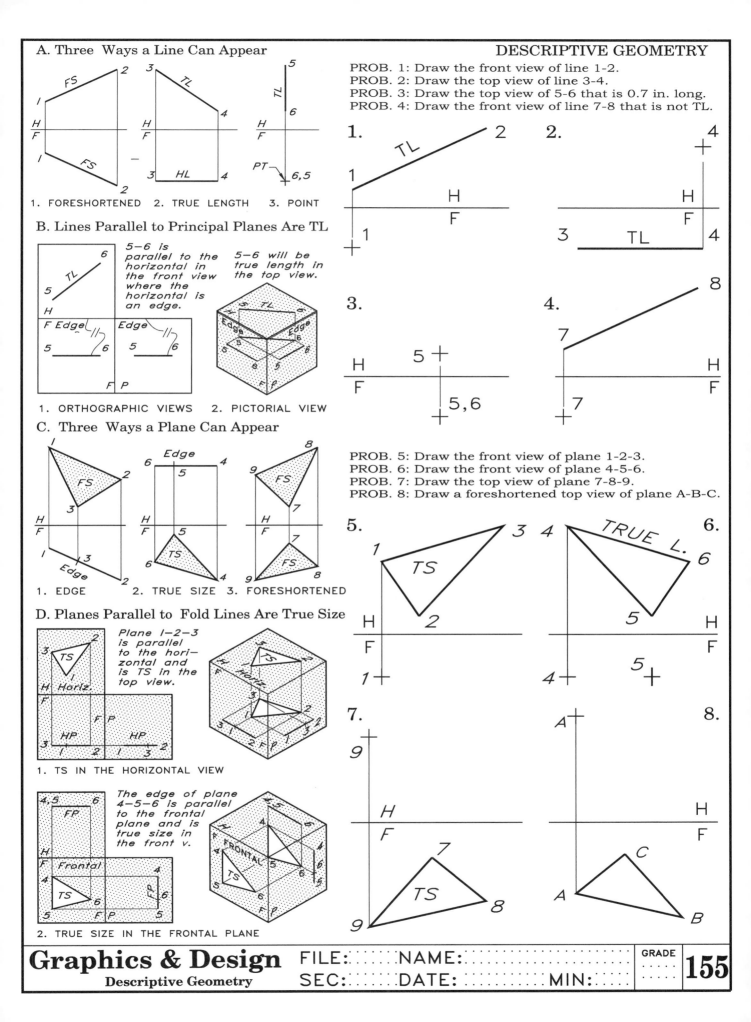

A. Three Ways a Line Can Appear

1. FORESHORTENED   2. TRUE LENGTH   3. POINT

B. Lines Parallel to Principal Planes Are TL

5-6 is parallel to the horizontal in the front view where the horizontal is an edge.

5-6 will be true length in the top view.

1. ORTHOGRAPHIC VIEWS   2. PICTORIAL VIEW

C. Three Ways a Plane Can Appear

1. EDGE   2. TRUE SIZE   3. FORESHORTENED

D. Planes Parallel to Fold Lines Are True Size

Plane 1-2-3 is parallel to the horizontal and is TS in the top view.

1. TS IN THE HORIZONTAL VIEW

The edge of plane 4-5-6 is parallel to the frontal plane and is true size in the front v.

2. TRUE SIZE IN THE FRONTAL PLANE

DESCRIPTIVE GEOMETRY

PROB. 1: Draw the front view of line 1-2.
PROB. 2: Draw the top view of line 3-4.
PROB. 3: Draw the top view of 5-6 that is 0.7 in. long.
PROB. 4: Draw the front view of line 7-8 that is not TL.

1.

2.

3.

4.

PROB. 5: Draw the front view of plane 1-2-3.
PROB. 6: Draw the front view of plane 4-5-6.
PROB. 7: Draw the top view of plane 7-8-9.
PROB. 8: Draw a foreshortened top view of plane A-B-C.

5.

6.

7.

8.

A. Locating a Point on a Plane

Project the line and O to the top view.

STEP 1 — Thru point O in any direction

STEP 2

B. Drawing a Line Parallel to Plane

Draw parallel to 3-4 thru point O.

Parallel

STEP 1 — Draw parallel to 3-4 in the front view.

STEP 2 — Parallel

C. A Line Parallel to a Plane

Parallel

GIVEN

SOLUTION

D. Drawing a Plane Parallel to a Plane

Parallel

STEP 1

STEP 2

STEP 1: Draw EF parallel to any line in the plane, 1-2 in this case, thru O in both views.
STEP 2: Draw GH through O parallel to 2-3 in both views.
Two intersecting lines form a plane like you did when you built a kite with two sticks.

PROBLEM 1: Draw a line thru point O that is parallel to line 1-2. Label both ends of the line when you have completed your construction.

PROBLEM 2: Draw a line thru point O that is parallel to plane 3-4-5 in both views. Label your construction when completed.

PROBLEM 3: Draw a plane thru point O that is parallel to plane 6-7-8. Label your construction when completed.

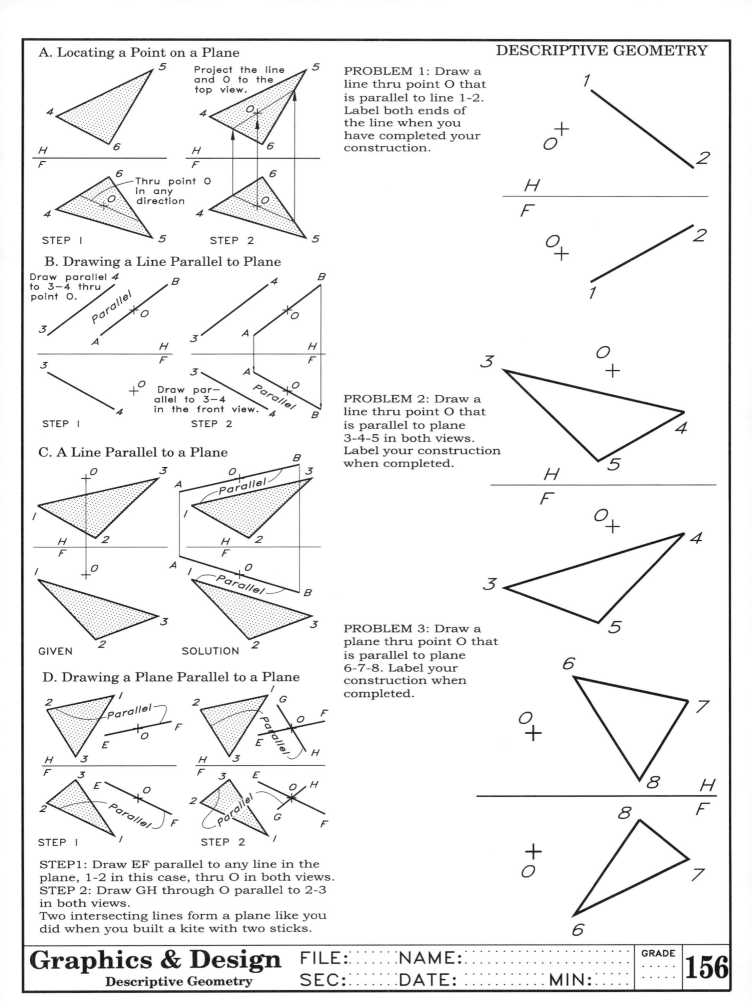

Graphics & Design

Assignment Solution

FILE: NAME:

SEC: DATE:

MIN:

A. True Length by an Auxiliary View

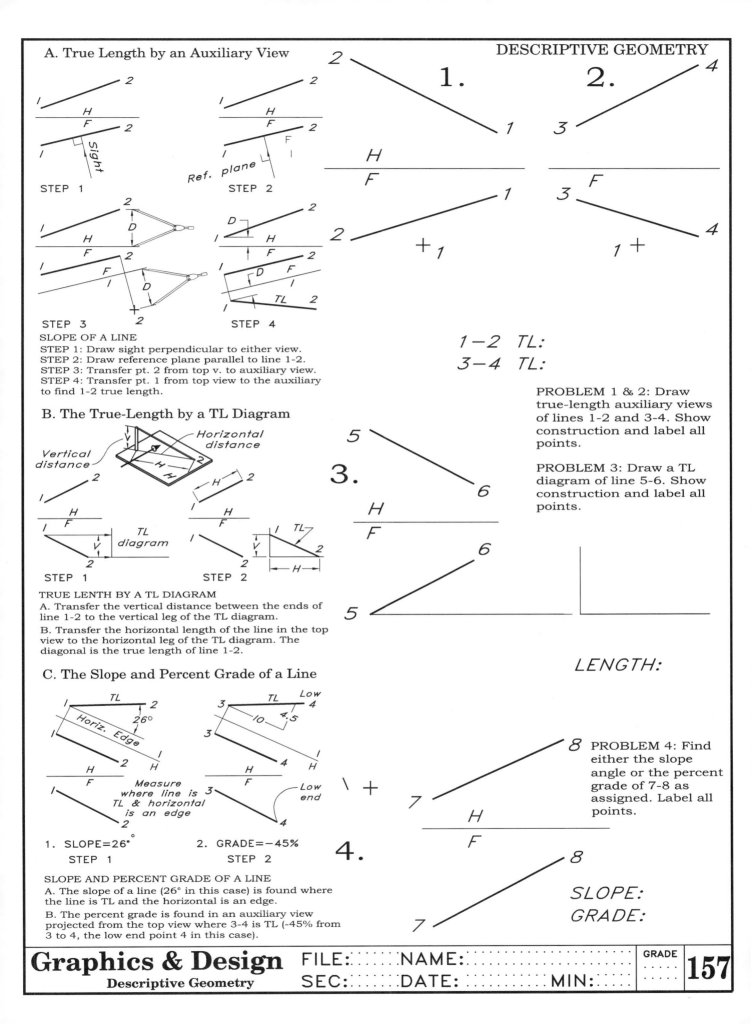

STEP 1

STEP 2

STEP 3

STEP 4

SLOPE OF A LINE
STEP 1: Draw sight perpendicular to either view.
STEP 2: Draw reference plane parallel to line 1-2.
STEP 3: Transfer pt. 2 from top v. to auxiliary view.
STEP 4: Transfer pt. 1 from top view to the auxiliary to find 1-2 true length.

B. The True-Length by a TL Diagram

STEP 1

STEP 2

TRUE LENTH BY A TL DIAGRAM
A. Transfer the vertical distance between the ends of line 1-2 to the vertical leg of the TL diagram.
B. Transfer the horizontal length of the line in the top view to the horizontal leg of the TL diagram. The diagonal is the true length of line 1-2.

C. The Slope and Percent Grade of a Line

1. SLOPE=26°
   STEP 1

2. GRADE=−45%
   STEP 2

SLOPE AND PERCENT GRADE OF A LINE
A. The slope of a line (26° in this case) is found where the line is TL and the horizontal is an edge.
B. The percent grade is found in an auxiliary view projected from the top view where 3-4 is TL (-45% from 3 to 4, the low end point 4 in this case).

1.

2.

1−2 TL:

3−4 TL:

PROBLEM 1 & 2: Draw true-length auxiliary views of lines 1-2 and 3-4. Show construction and label all points.

PROBLEM 3: Draw a TL diagram of line 5-6. Show construction and label all points.

3.

LENGTH:

PROBLEM 4: Find either the slope angle or the percent grade of 7-8 as assigned. Label all points.

4.

SLOPE:
GRADE:

Graphics & Design
Descriptive Geometry

FILE: . . . . . . NAME: . . . .

SEC: . . . . . DATE: . . . . . . MIN: . . . .

GRADE

157

6.4,6.4

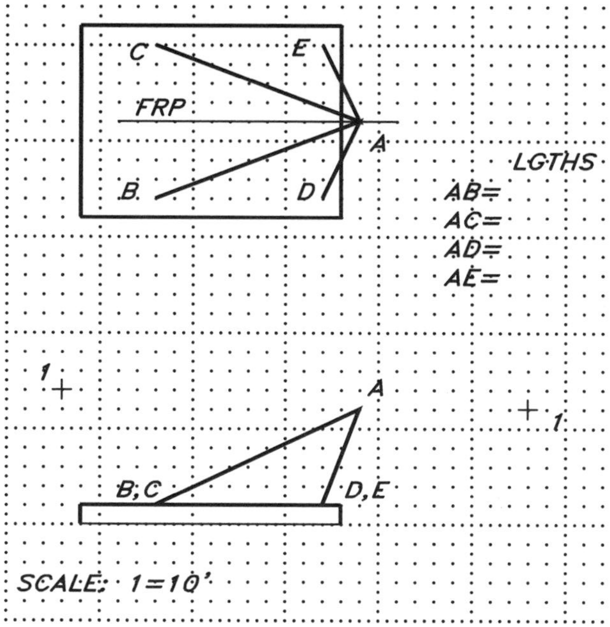

LGTHS

AB=
AC=
AD=
AE=

C
E
A
FRP
D
B

A
D,E
B,C

+1
1+

SCALE: 1=10'

0,0

TRUE-LENGTH VIEWS

BY AUXILIARY VIEWS, FIND THE TRUE-
LENGTH VIEWS OF THE STRUCTURAL
MEMBERS USED FOR LIFTING CONCRETE
SLABS INTO AN UPRIGHT POSITION.

Graphics & Design
Descriptive Geometry

FILE: .... NAME: ....
SEC: .... DATE: ....

MIN: ....

A. Edge View of a Plane

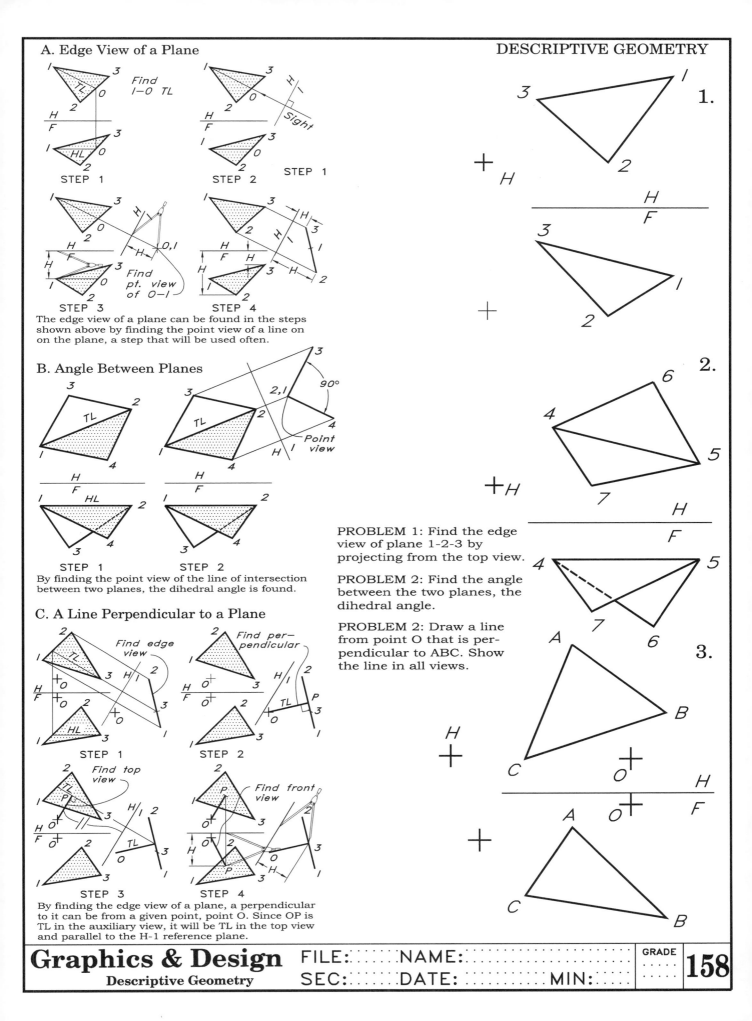

STEP 1   STEP 2   STEP 1

Find 1–0 TL

STEP 3   STEP 4

Find pt. view of 0–1

The edge view of a plane can be found in the steps shown above by finding the point view of a line on on the plane, a step that will be used often.

B. Angle Between Planes

STEP 1   STEP 2

By finding the point view of the line of intersection between two planes, the dihedral angle is found.

C. A Line Perpendicular to a Plane

STEP 1   STEP 2

STEP 3   STEP 4

By finding the edge view of a plane, a perpendicular to it can be from a given point, point O. Since OP is TL in the auxiliary view, it will be TL in the top view and parallel to the H-1 reference plane.

1.

2.

PROBLEM 1: Find the edge view of plane 1-2-3 by projecting from the top view.

PROBLEM 2: Find the angle between the two planes, the dihedral angle.

PROBLEM 2: Draw a line from point O that is per-pendicular to ABC. Show the line in all views.

3.

Graphics & Design
Descriptive Geometry

Graphics & Design

Assignment Solution

GRADE

FILE: NAME:

SEC: DATE: MIN:

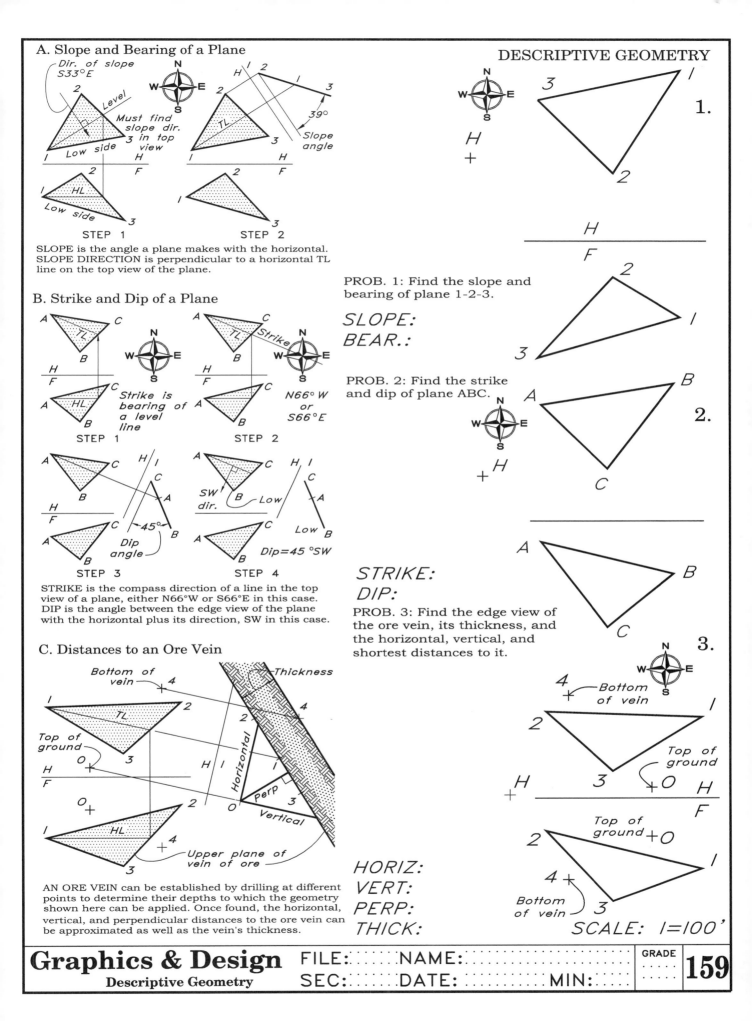

A. Slope and Bearing of a Plane

Dir. of slope S33°E

Level

Must find slope dir. 3 in top view

Low side

STEP 1

STEP 2

39°
Slope angle

SLOPE is the angle a plane makes with the horizontal.
SLOPE DIRECTION is perpendicular to a horizontal TL line on the top view of the plane.

B. Strike and Dip of a Plane

STEP 1

Strike is bearing of a level line

STEP 2

N66°W or S66°E

STEP 3

45°
Dip angle

STEP 4

SW dir.
Low
Low
Dip=45°SW

STRIKE is the compass direction of a line in the top view of a plane, either N66°W or S66°E in this case.
DIP is the angle between the edge view of the plane with the horizontal plus its direction, SW in this case.

C. Distances to an Ore Vein

Bottom of vein
Thickness
Top of ground
Horizontal
Perp
Vertical
Upper plane of vein of ore

AN ORE VEIN can be established by drilling at different points to determine their depths to which the geometry shown here can be applied. Once found, the horizontal, vertical, and perpendicular distances to the ore vein can be approximated as well as the vein's thickness.

DESCRIPTIVE GEOMETRY

1.

PROB. 1: Find the slope and bearing of plane 1-2-3.

SLOPE:
BEAR.:

PROB. 2: Find the strike and dip of plane ABC.

2.

STRIKE:
DIP:

PROB. 3: Find the edge view of the ore vein, its thickness, and the horizontal, vertical, and shortest distances to it.

3.

Bottom of vein
Top of ground

Top of ground +O
Bottom of vein

HORIZ:
VERT:
PERP:
THICK:

SCALE: 1=100'

Graphics & Design
Descriptive Geometry

FILE: NAME:
SEC: DATE: MIN:

GRADE

159

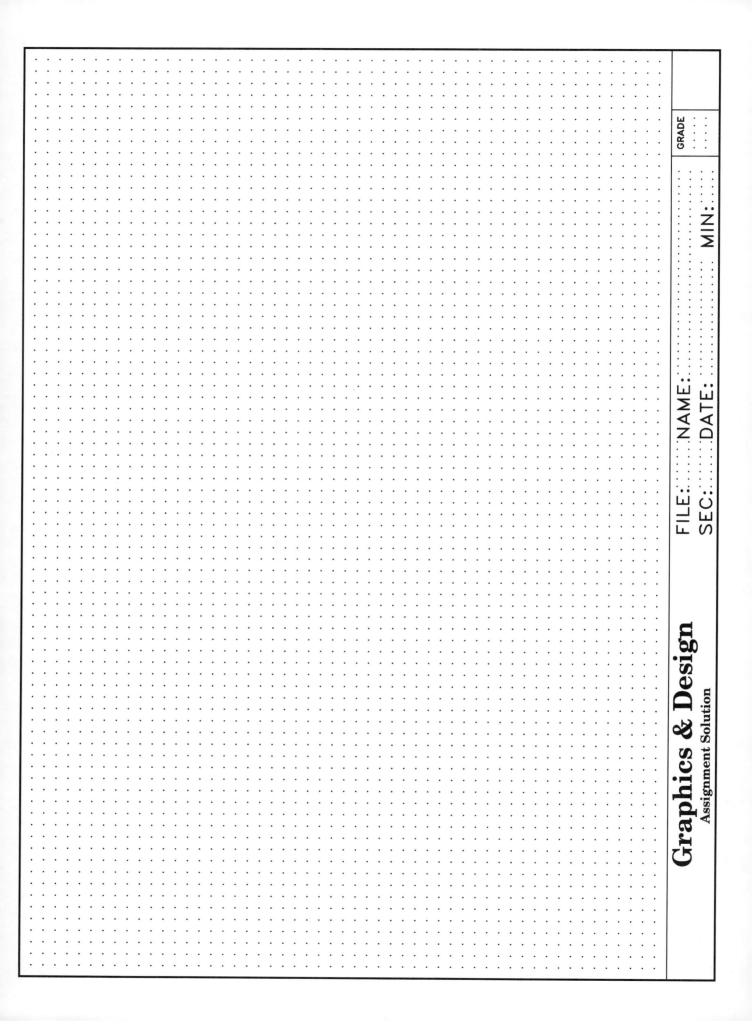

Graphics & Design

Assignment Solution

FILE: _____ NAME: _____

SEC: _____ DATE: _____

GRADE

MIN: _____

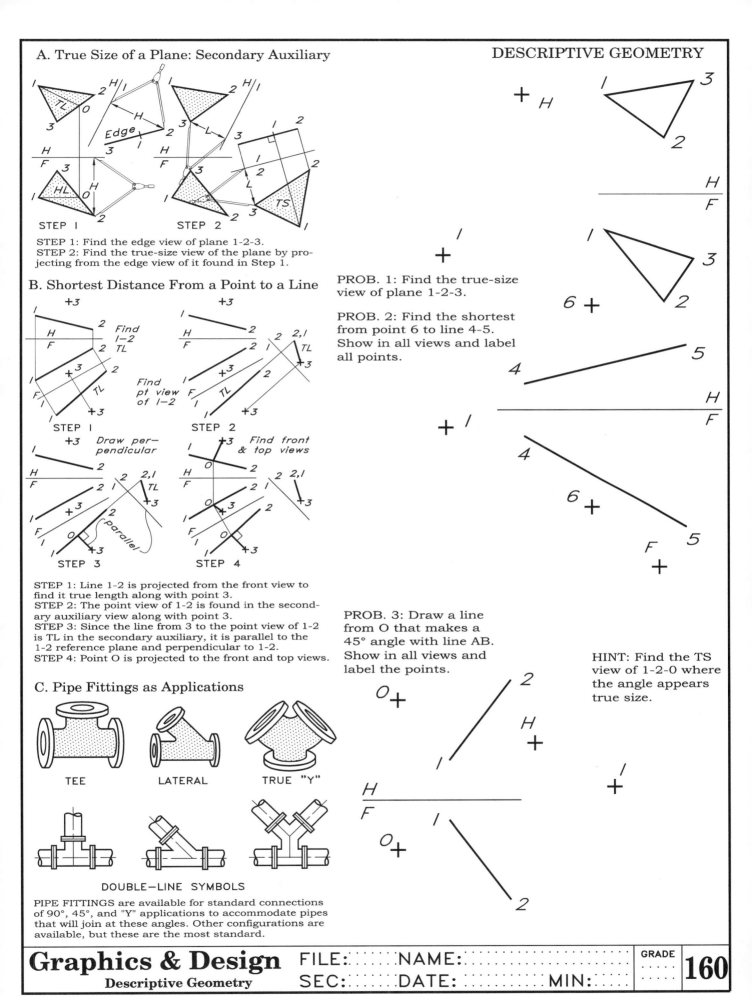

A. True Size of a Plane: Secondary Auxiliary

STEP 1   STEP 2

STEP 1: Find the edge view of plane 1-2-3.
STEP 2: Find the true-size view of the plane by projecting from the edge view of it found in Step 1.

B. Shortest Distance From a Point to a Line

STEP 1   STEP 2

STEP 3   STEP 4

STEP 1: Line 1-2 is projected from the front view to
find it true length along with point 3.
STEP 2: The point view of 1-2 is found in the secondary auxiliary view along with point 3.
STEP 3: Since the line from 3 to the point view of 1-2
is TL in the secondary auxiliary, it is parallel to the
1-2 reference plane and perpendicular to 1-2.
STEP 4: Point O is projected to the front and top views.

C. Pipe Fittings as Applications

TEE   LATERAL   TRUE "Y"

DOUBLE—LINE SYMBOLS

PIPE FITTINGS are available for standard connections
of 90°, 45°, and "Y" applications to accommodate pipes
that will join at these angles. Other configurations are
available, but these are the most standard.

PROB. 1: Find the true-size
view of plane 1-2-3.

PROB. 2: Find the shortest
from point 6 to line 4-5.
Show in all views and label
all points.

PROB. 3: Draw a line
from O that makes a
45° angle with line AB.
Show in all views and
label the points.

HINT: Find the TS
view of 1-2-0 where
the angle appears
true size.

Graphics & Design
Descriptive Geometry

FILE:........NAME:............................
SEC:........DATE:..........MIN:..........

GRADE

160

Graphics & Design

Assignment Solution

FILE: ⋯⋯ NAME: ⋯⋯⋯
SEC: ⋯⋯ DATE: ⋯⋯⋯
MIN:

GRADE

DESIGN APPLICATIONS should be the most fun of this course because the problems are similar to those encountered in industry.

Lay out the necessary orthographic views to determine the specified design information. Problems can be drawn either on A- or B-size sheets. (Most of these problems have been adapted from those encountered at the Boeing Company.)

DESIGN: LIFTING HOOK
Determine the following geometry: The length of the cylindrical rod and its bend angle.

Solve on a size A sheet.

Weight: 56 lbs

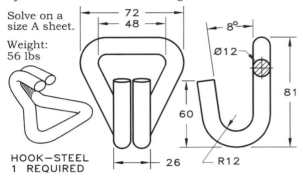

72
48
8°
Ø12
81
60
26
R12

HOOK—STEEL
1 REQUIRED

DESIGN: CLEARANCE
Determine the clearance betweeen the web and the tube.

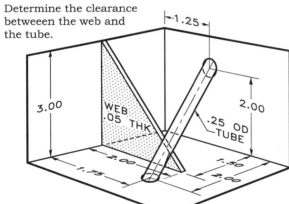

1.25
3.00
2.00
WEB .05 THK
.25 OD TUBE
2.00
1.75
1.50
2.00

ANALYSIS: RIGID TUBE
A 0.50 dia. tube connects A with D with two bends of radii of 2 in. at the centerline at B and C. Determine the angles and length from A to D.

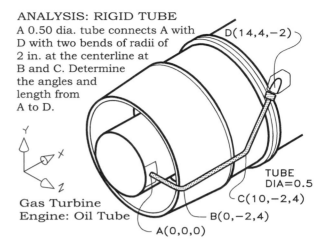

D(14,4,−2)

Gas Turbine Engine: Oil Tube

TUBE DIA=0.5
C(10,−2,4)
B(0,−2,4)
A(0,0,0)

DESIGN: TABLE SUPPORT
The frame supports an 18X48 glass top with Ø.34 chromed steel tubes. The design is simple but the geometry is not.

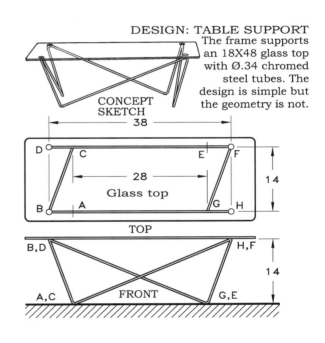

CONCEPT SKETCH

38
D C E F
28
Glass top
14
B A G H

TOP

B,D H,F
14
A,C FRONT G,E

DESIGN: BRACKET
Determine the required geometry and details.

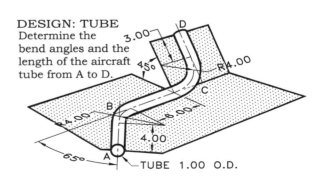

.16 THICK
TRUE BEND ANGLE, NOT TS IN FRONT V.
2.30
3.00
45°
3.00
60°
45°
6.00
INSIDE FILLETS R.30

(16) CONNECTOR STRAP
1020 STEEL—2 REQUIRED

DESIGN: TUBE
Determine the bend angles and the length of the aircraft tube from A to D.

3.00
45°
R4.00
D
C
R4.00
B
8.00
4.00
65°
A
TUBE 1.00 O.D.

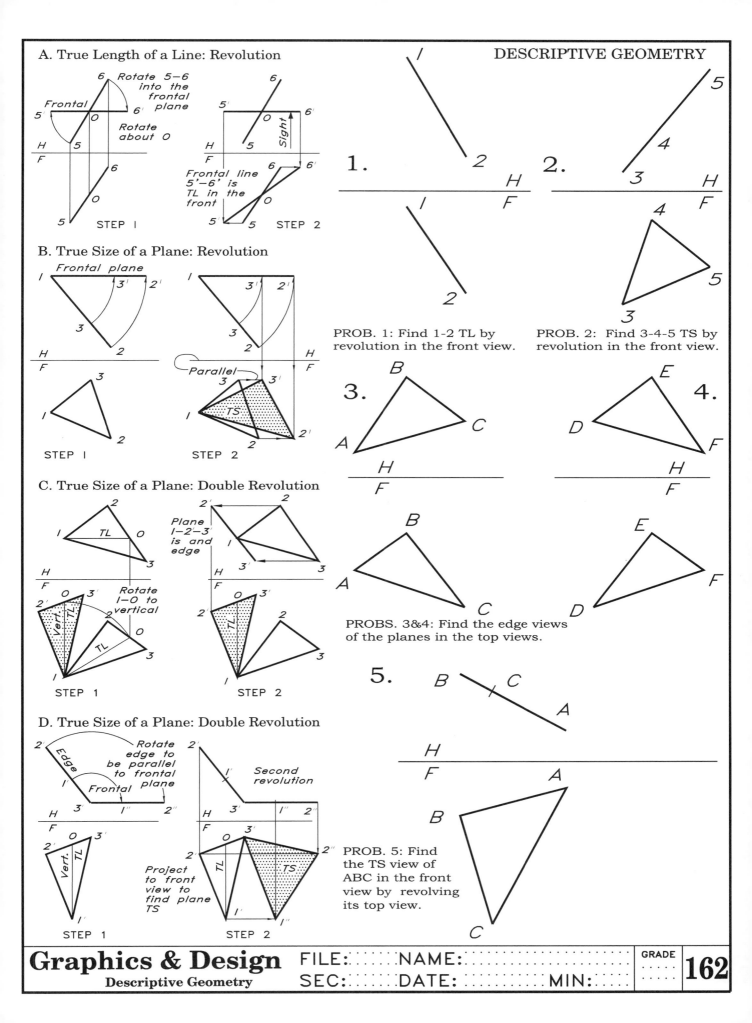

A. True Length of a Line: Revolution

Rotate 5–6 into the frontal plane

Frontal

Rotate about O

STEP 1

Sight

Frontal line 5'–6' is TL in the front

STEP 2

B. True Size of a Plane: Revolution

Frontal plane

Parallel

TS

STEP 1 STEP 2

C. True Size of a Plane: Double Revolution

TL

Plane 1–2'–3' is and edge

Rotate 1–O to vertical

Vert. TL

TL

STEP 1 STEP 2

D. True Size of a Plane: Double Revolution

Edge

Rotate edge to be parallel to frontal plane

Frontal

Second revolution

Vert. TL

Project to front view to find plane TS

TL

TS

STEP 1 STEP 2

DESCRIPTIVE GEOMETRY

1.

H
F

2.

H
F

PROB. 1: Find 1-2 TL by revolution in the front view.

PROB. 2: Find 3-4-5 TS by revolution in the front view.

3.

B
A
C

H
F

4.

E
D
F

H
F

B
A
C

E
D
F

PROBS. 3&4: Find the edge views of the planes in the top views.

5.

B C
A

H
F

A
B

C

PROB. 5: Find the TS view of ABC in the front view by revolving its top view.

Graphics & Design
Descriptive Geometry

FILE: NAME:
SEC: DATE: MIN:

GRADE

162

A. Coplanar Forces in Equilibrium

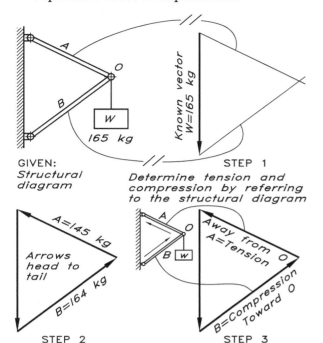

GIVEN:
Structural diagram

Arrows head to tail

A=145 kg
B=164 kg

STEP 2

Determine tension and compression by referring to the structural diagram

*Away from O
A=Tension*

*B=Compression
Toward O*

STEP 3

STEP 1: Draw vector of 165 kg true length and vectors A and B parallel to their directions at each end.
STEP 2: Draw vectors head-to-tail beginning with the only known vector, the 165 kg load.
STEP 3: Vector A points away from O and is in tension. Vector B points toward O and is in compression.

B. Coplanar Forces with a Pulley

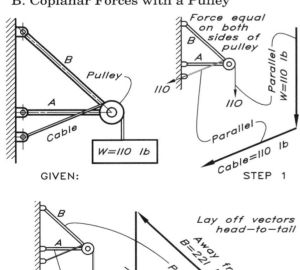

GIVEN:

Force equal on both sides of pulley

Parallel
W=110 lb
110
110
Parallel
Cable=110 lb

STEP 1

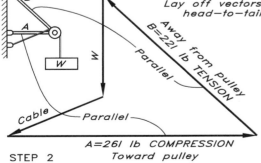

Lay off vectors head-to-tail

*Away from pulley
B=221 lb TENSION*

Parallel

Cable

Parallel

A=261 lb COMPRESSION
Toward pulley

STEP 2

STEP 1: Draw the cable vectors head to tail and parallel to their structural members.
STEP 2: Draw vectors A and B to close the polygon, giving the loads in each member. A is in compression and B is in tension.

DESCRIPTIVE GEOMETRY: VECTORS

PROBLEM 1: The truss is to be loaded with 300 lbs as shown. Determine the loads that must be supported by members A and B and indicate whether these loads are in tension or compression.

SCALE: 1=100 LBS

A
O
B
W
300 LBS

W=
300 LBS

PROBLEM 2: The truss is to be loaded on a cable with 175 lbs as shown. Determine the loads that will be supported by members A and B and specify tension or compression

*SCALE:
1=100LBS*

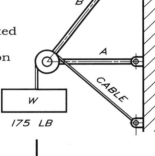

B
A
CABLE
W
175 LB

175 LBS
LOAD

Graphics & Design
Descriptive Geometry: Vectors
FILE: NAME:
SEC: DATE: MIN:
GRADE
163

Graphics & Design
Assignment Solution

FILE: NAME:
SEC: DATE: MIN:

GRADE

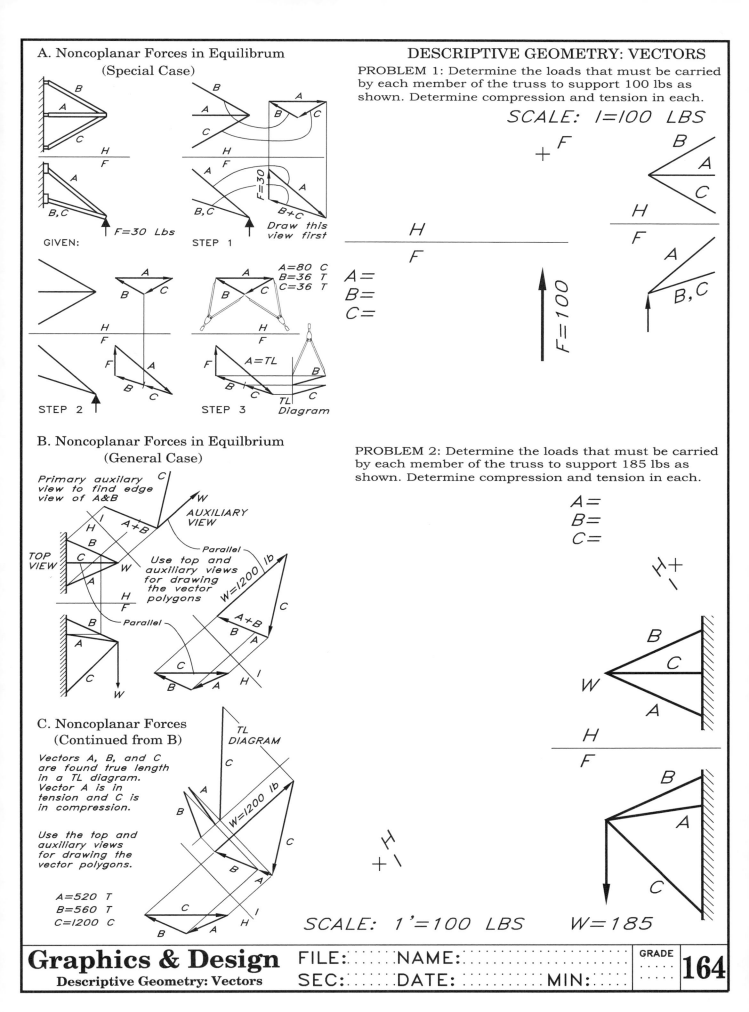

A. Noncoplanar Forces in Equilibrum
(Special Case)

GIVEN:

F=30 Lbs

STEP 1

Draw this view first

F=30

A=80 C
B=36 T
C=36 T

STEP 2

STEP 3

A=TL

TL Diagram

B. Noncoplanar Forces in Equilbrium
(General Case)

Primary auxilary view to find edge view of A&B

AUXILIARY VIEW

TOP VIEW

Use top and auxiliary views for drawing the vector polygons

Parallel

Parallel

W=1200 lb

C. Noncoplanar Forces
(Continued from B)

Vectors A, B, and C are found true length in a TL diagram. Vector A is in tension and C is in compression.

Use the top and auxiliary views for drawing the vector polygons.

A=520 T
B=560 T
C=1200 C

TL DIAGRAM

W=1200 lb

DESCRIPTIVE GEOMETRY: VECTORS

PROBLEM 1: Determine the loads that must be carried by each member of the truss to support 100 lbs as shown. Determine compression and tension in each.

SCALE: 1=100 LBS

+ F

B
A
C

H
F

A
B,C

A=
B=
C=

F=100

PROBLEM 2: Determine the loads that must be carried by each member of the truss to support 185 lbs as shown. Determine compression and tension in each.

A=
B=
C=

B
C
W
A

H
F

B
A
C

SCALE: 1'=100 LBS

W=185

Graphics & Design
Descriptive Geometry: Vectors

FILE: ⋯ NAME: ⋯
SEC: ⋯ DATE: ⋯ MIN: ⋯

GRADE

164

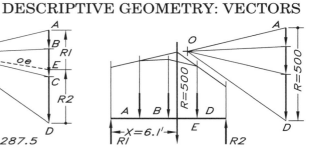

Vector Diagram

100 150 250

Space Diagram

STEP 1: Letter the spaces between the loads using Bow's notation. Draw the vertical loads head-to-tail in a vector diagram. Locate pole point O at a convenient location and draw strings from O to the ends of each vector.

Funicular Diagram

R1 212.5   R2 287.5

STEP 2: Extend the lines of the vertical loads and draw a funicular diagram with string oa in the A space, ob in the B space, oc in the C space, and so on. The last string, oe, closes the diagram. Transfer oe to the vector diagram to locate E, thus establishing R1 and R2 which are EA and DE,

X=6.1'

R=500

STEP 3: The resultant of the three downward forces equals their graphical summation, line AD. Locate the resultant by extending oa and od in the funicular diagram to their intersection. The resultant, R=500 lb, acts through this point in a downward direction at X=6.1 ft. from the left

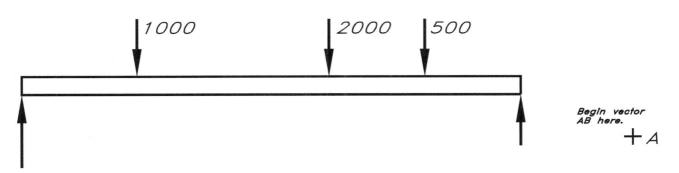

1000   2000   500

Begin vector AB here.

+A

SCALE: 1=10 FT

O +

PROBLEM: You are required to find the loads in the support members at each end of the beam and the location and magnitude of the single resultant of the three loads given at A, B, and C. Use Bow's notation to label the spaces between the vectors in a clockwise For example, the first vector is labeled A on one side B on the other and is referred to as AB. The next vector to its right is labeled BC and will join with vector AB to begin the formation of the vector diagram and the funicular diagram.

SCALE: 1=1000 LB

Graphics & Design
Descriptive Geometry: Vectors

FILE: NAME:
SEC: DATE: MIN:

GRADE

165

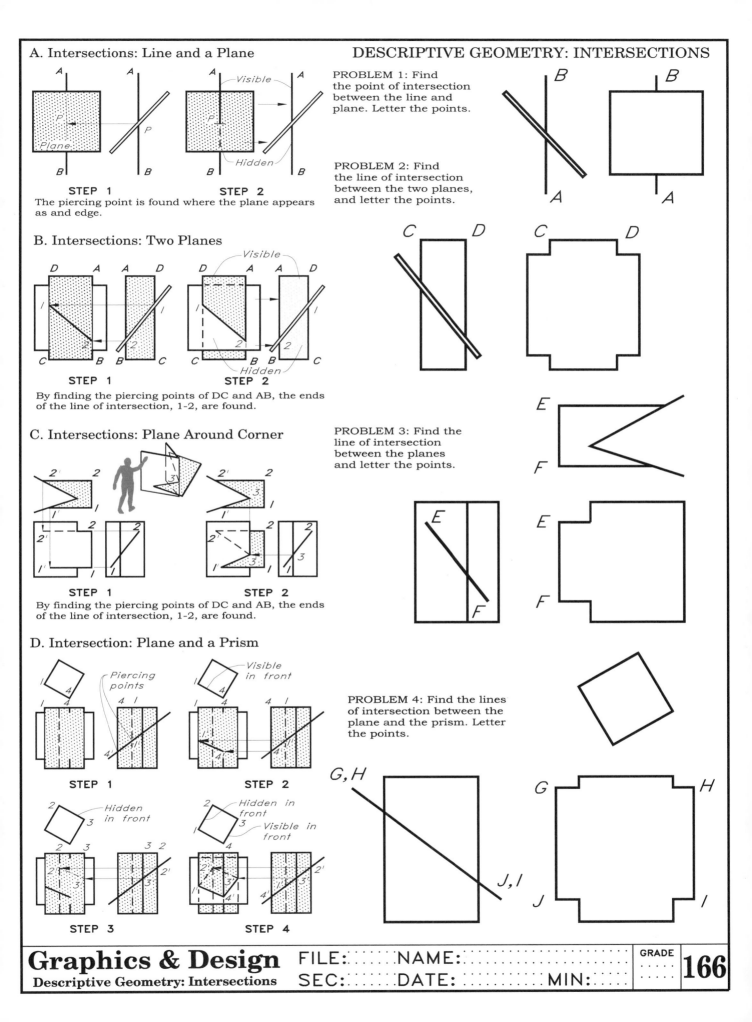

A. Intersections: Line and a Plane

STEP 1 **STEP 2**

The piercing point is found where the plane appears as and edge.

B. Intersections: Two Planes

Visible

Hidden

STEP 1 **STEP 2**

By finding the piercing points of DC and AB, the ends of the line of intersection, 1-2, are found.

C. Intersections: Plane Around Corner

STEP 1 **STEP 2**

By finding the piercing points of DC and AB, the ends of the line of intersection, 1-2, are found.

D. Intersection: Plane and a Prism

Piercing points

Visible in front

STEP 1 **STEP 2**

Hidden in front

Hidden in front

Visible in front

STEP 3 **STEP 4**

DESCRIPTIVE GEOMETRY: INTERSECTIONS

PROBLEM 1: Find the point of intersection between the line and plane. Letter the points.

PROBLEM 2: Find the line of intersection between the two planes, and letter the points.

PROBLEM 3: Find the line of intersection between the planes and letter the points.

PROBLEM 4: Find the lines of intersection between the plane and the prism. Letter the points.

A. Intersections: Two Prisms

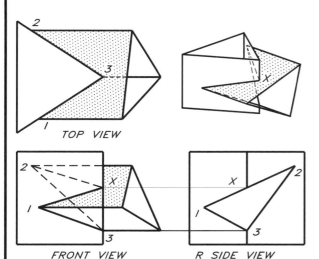

TOP VIEW

FRONT VIEW   R SIDE VIEW

The intersections of planes is best found when one of of the planes appears as and edge as in the side view.

B. Intersection by Projection: Two Prisms

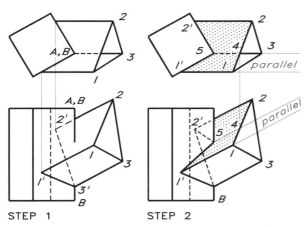

STEP 1   STEP 2

Intersections can also be found by projection when a plane does not appear as an edge as shown here.

An auxiliary view is also helpful in finding intersections between planes as shown in the auxiliary view.

C. Intersection by an Auxiliary View

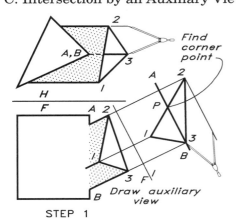

Find corner point

Draw auxiliary view

STEP 1

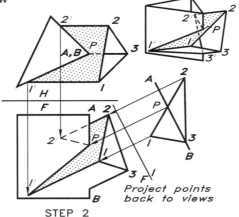

Project points back to views

STEP 2

Complete views

STEP 3

DESCRIPTIVE GEOMETRY: INTERSECTIONS

PROBLEM 1: Find the intersection of the two prisms, show visibility, and label the points.

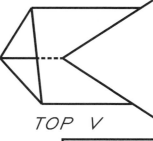

TOP V

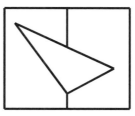

L SIDE VIEW

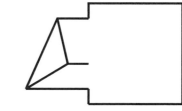

FRONT VIEW

PROBLEM 2: Find the intersection between the prisms by using either of the two methods shown as assigned. Label the points and show visibility.

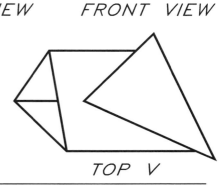

TOP V

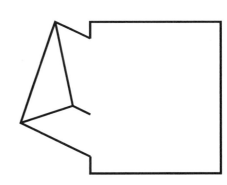

FRONT V

Graphics & Design
Descriptive Geometry: Intersections

A. Intersections: Inclined Cylinders

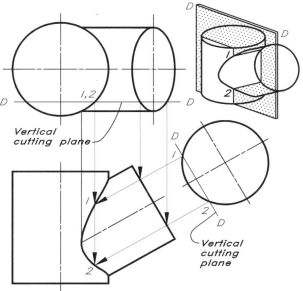

Vertical cutting plane

Vertical cutting plane

A series of of frontal cutting planes parallel to plane D-D are used to locate the line of intersection in the front view.

B. Intersections: Plane and Cylinder

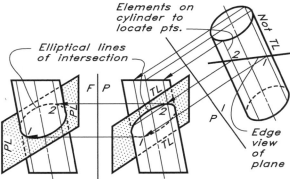

Elements on cylinder to locate pts.

Elliptical lines of intersection

The edge view of the plane is drawn in the auxiliary view to locate a series of intersecting points on the cylinder.

C. Intersections: Pyramid and Prism

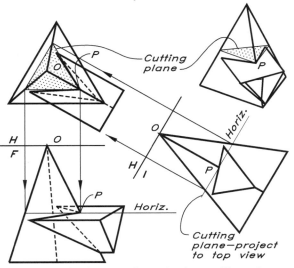

Cutting plane

Horiz.

Horiz.

Cutting plane—project to top view

Horizontal reference planes in the auxiliary view are used to locate piercing points in the top and front views.

DESCRIPTIVE GEOMETRY: INTERSECTIONS

PROBLEM 1: Find the line of intersection between the two cylinders and label points.

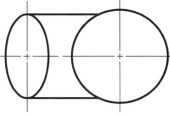

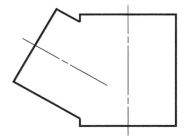

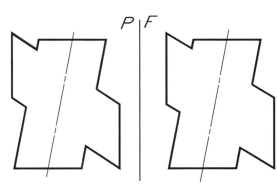

PROBLEM 2: Find the line of of intersection between the cylinder and plane and label the points.

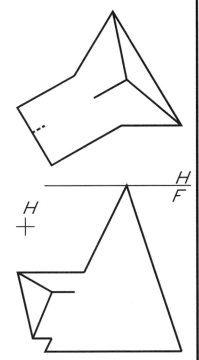

PROBLEM 3: Find the line of intersection between the prism and pyramid. Label the points.

Graphics & Design
Descriptive Geometry: Intersections

FILE: .... NAME: ....

SEC: .... DATE: .... MIN: ....

GRADE

168

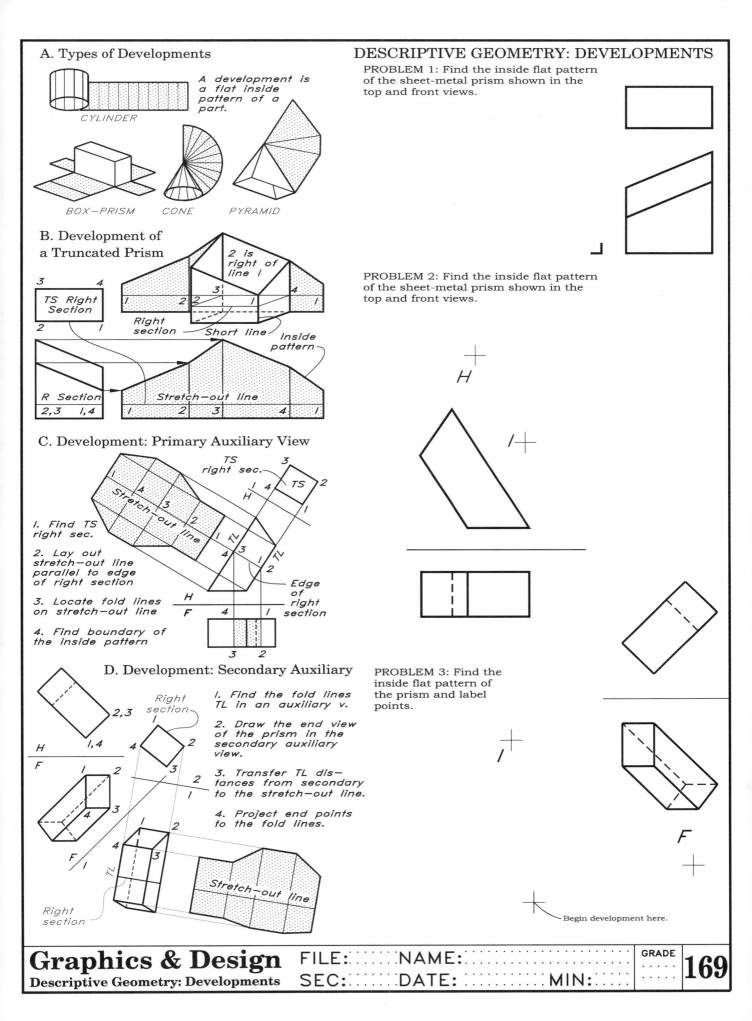

A. Types of Developments

A development is a flat inside pattern of a part.

CYLINDER

BOX—PRISM   CONE   PYRAMID

B. Development of a Truncated Prism

3   4
TS Right Section
2   1

2 is right of line 1

Right section   Short line   Inside pattern

R Section
2,3 1,4

Stretch-out line
1   2   3   4   1

C. Development: Primary Auxiliary View

TS right sec.

Stretch-out line

1. Find TS right sec.

2. Lay out stretch—out line parallel to edge of right section

3. Locate fold lines on stretch—out line

4. Find boundary of the inside pattern

Edge of right section

H
F   4   1

3   2

D. Development: Secondary Auxiliary

Right section

H
F   1,4

1. Find the fold lines TL in an auxiliary v.

2. Draw the end view of the prism in the secondary auxiliary view.

3. Transfer TL distances from secondary to the stretch—out line.

4. Project end points to the fold lines.

Stretch-out line

Right section

PROBLEM 1: Find the inside flat pattern of the sheet-metal prism shown in the top and front views.

PROBLEM 2: Find the inside flat pattern of the sheet-metal prism shown in the top and front views.

H

F

PROBLEM 3: Find the inside flat pattern of the prism and label points.

F

Begin development here.

Graphics & Design
Descriptive Geometry: Developments

FILE: .... NAME: ....
SEC: .... DATE: .... MIN: ....

GRADE

169

A. Development: A Truncated Cylinder

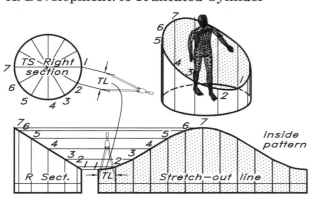

Only true-length lines can be used to construct developments which makes this problem ideal for drawing this cylindrical part.

B. Development: Cylinder With a Secondary Auxiliary

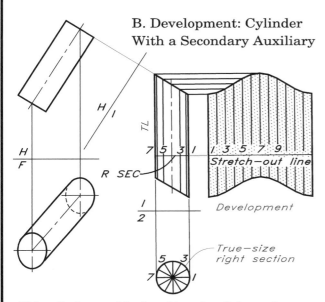

This cylinder must be found true length in a primary auxiliary view and as a true circle in a secondary auxiliary view.

C. Development: A Pyramid

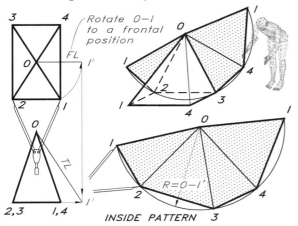

A right pyramid will have its points located on the arc formed by the true length of its diagonal corners and true-lengths of the horizontal base line in the top view.

PROBLEM 1: Find the inside developed flat development of the cylinder given here. Label your points of construction.

PROBLEM 2: Find the inside flat pattern of the sheet-metal cylinder shown in the top and front views.

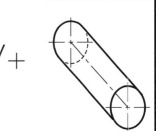

PROBLEM 3: Find the inside flat pattern of the sheet-metal prism shown in the top and front views. Label the points and show your construction.

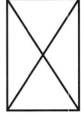

Graphics & Design
Descriptive Geometry: Developments

FILE: .... NAME: ....

SEC: .... DATE: .... MIN: ......

GRADE **170**

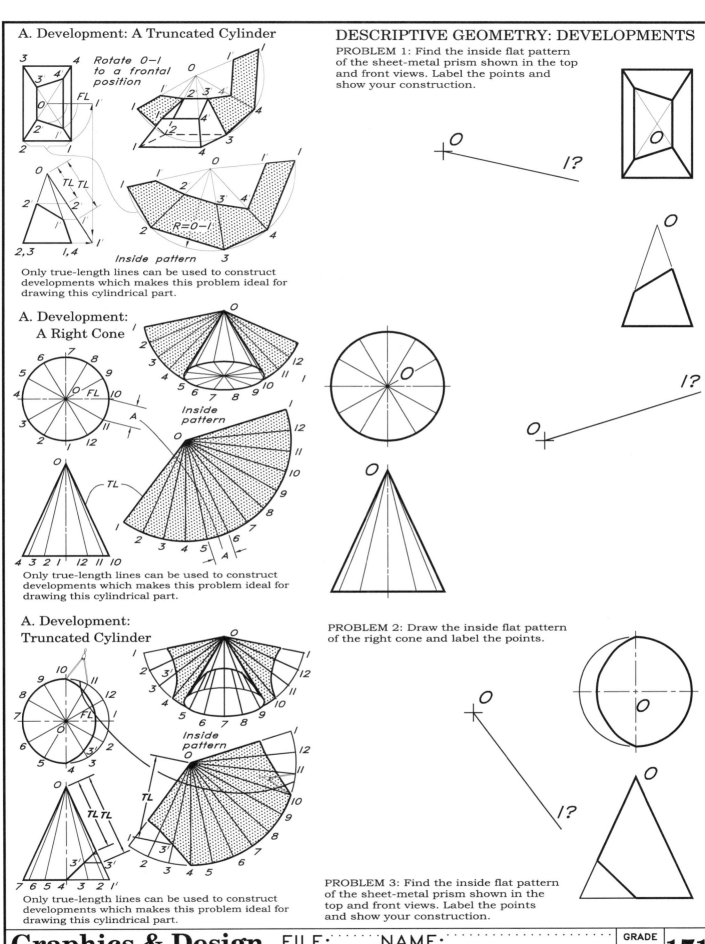

A. Development: A Truncated Cylinder

Rotate 0–1 to a frontal position

Only true-length lines can be used to construct developments which makes this problem ideal for drawing this cylindrical part.

A. Development: A Right Cone

Inside pattern

Only true-length lines can be used to construct developments which makes this problem ideal for drawing this cylindrical part.

A. Development: Truncated Cylinder

Inside pattern

Only true-length lines can be used to construct developments which makes this problem ideal for drawing this cylindrical part.

DESCRIPTIVE GEOMETRY: DEVELOPMENTS

PROBLEM 1: Find the inside flat pattern of the sheet-metal prism shown in the top and front views. Label the points and show your construction.

PROBLEM 2: Draw the inside flat pattern of the right cone and label the points.

PROBLEM 3: Find the inside flat pattern of the sheet-metal prism shown in the top and front views. Label the points and show your construction.

Graphics & Design
Descriptive Geometry: Developments

FILE: . . . . . . NAME: . . . .
SEC: . . . . . DATE: . . . . MIN: . . . . . .

GRADE

171